Introduction of SUPER-SPEED RAIL

Introduction of
SUPER-SPEED RAIL

Qizhou Hu
Nanjing University of Science and Technology, China

World Scientific

NEW JERSEY · LONDON · SINGAPORE · BEIJING · SHANGHAI · HONG KONG · TAIPEI · CHENNAI · TOKYO

Published by

World Scientific Publishing Co. Pte. Ltd.

5 Toh Tuck Link, Singapore 596224

USA office: 27 Warren Street, Suite 401-402, Hackensack, NJ 07601

UK office: 57 Shelton Street, Covent Garden, London WC2H 9HE

Library of Congress Cataloging-in-Publication Data

Names: Hu, Qizhou, author.

Title: Introduction of super-speed rail / Qizhou Hu,
    Nanjing University of Science and Technology, China.

Description: New Jersey : World Scientific, [2023] | Includes bibliographical references and index.

Identifiers: LCCN 2022058164 | ISBN 9789811270093 (hardcover) |
    ISBN 9789811270109 (ebook for institutions) | ISBN 9789811270116 (ebook for individuals)

Subjects: LCSH: High speed trains. | Pneumatic-tube transportation. | Vacuum technology.

Classification: LCC TF1620 .H83 2023 | DDC 625.26/5--dc23/eng/20230111

LC record available at https://lccn.loc.gov/2022058164

British Library Cataloguing-in-Publication Data

A catalogue record for this book is available from the British Library.

For any available supplementary material, please visit
https://www.worldscientific.com/worldscibooks/10.1142/13242#t=suppl

Desk Editors: Soundararajan/Steven Patt

Typeset by Stallion Press
Email: enquiries@stallionpress.com

Preface

Super-Speed Rail (SSR) is a kind of vehicle designed on the theory of "Evacuated Tube Transport." It has the characteristics of ultra-high speed, high safety, low energy consumption, low noise, no vibration, and no pollution. SSR has the potential to be the next generation of transportation after cars, ships, trains, and planes, which is the fifth form of transportation. This book analyzes the SSR operation principle, system architecture, and attribute characteristics, discusses its feasibility, and analyzes the global integration issues in the SSR environment.

The advent of the SSR system is an upgrade of the traditional rail transit design concept, as well as a test of aerodynamic knowledge. The SSR system has faced a lot of controversies because it is so far ahead of its time, but technology is constantly improving. In the near future, the idea of SSR will no longer be a fantasy, but a vehicle that can be tested. This book was written by Hu Qizhou's team, mainly composed of Wu Xiaoyu, Li Chenyang, Wu Yikai, Fang Xin, QIU Lexia HONG Tai, HE Jun, LEI Aiguo, LEI Aiguo, and others.

As a Chinese saying goes, "benefit the world, shine on people's livelihood." As a popular science book, popularizing knowledge to the public is the highest goal of each of us. Some pictures and contents of this book are from the Internet. Because we cannot find the source of some information, we can only express our gratitude and respect here. We would like to express our heartfelt thanks to our colleagues in the editorial department for their selfless help in writing this book. The book is easy to read and understand, and is equipped with rich illustrations. It is suitable for high-speed rail enthusiasts, as well as scientific researchers, engineer technicians, management workers, teachers, and students of colleges and universities.

Due to the limited level of knowledge, there are inevitably omissions and mistakes in the book. Please give us your advice and criticism.

About the Author

Hu Qizhou, Ph.D. Supervisor, director of the Institute of High-speed Rail Science of Nanjing University of Science and Technology, distinguished Professor of Henan Polytechnic University, young and middle-aged academic leader of the "Blue Project" in Jiangsu Province, winner of the "Six Talent Peaks" high-level talent project in Jiangsu Province, and editor of the European journal "International Journal for Traffic and Transport Engineering." In recent years, he has undertaken more than 30 national and provincial-level scientific research projects, published more than 50 academic papers, published 7 monographs, edited 3 textbooks, 5 popular science books, and obtained 15 invention patents. 10 provincial and ministerial awards.

Contents

Chapter 1

Introduction

On October 1, 1964, the world's first high-speed railway (HSR) was operated in Japan, marking the birth of the HSR (Figure 1.1). Since then, the first round of the "HSR boom" was set off all over the world. However, due to technical limitations, the HSR was not vigorously developed, and the operation speed was less than 300 km/h.

On September 22, 1983, the operation of the first HSR line from Paris to Lyon in France (Figure 1.2) set off the second round of the "HSR boom" in the world. However, due to economic depression in various countries in the world, especially the limited economic capacity of developing countries, HSR was only built and operated in the economically developed countries.

On June 2, 1991, the operation of Germany's first HSR line from Mannheim to Stuttgart set off the third round of the "HSR boom" all over the world (Figure 1.3). However, due to a major HSR accident in Germany in 1998, countries around the world expressed a wait-and-see attitude toward HSR based on safety considerations, and did not vigorously promote and build it.

On August 1, 2008, China's Beijing–Tianjin Intercity Railway (Figure 1.4) with a design speed of 350 km/h was put into operation, marking China's entry into the era of HSR and setting off the fourth round of the "HSR boom" worldwide. Due to the rapid development of the HSR in China, HSR technology has become mature, economical, applicable, and reliable. HSR has been promoted rapidly all over the world.

Today, nine European countries with the HSR plan to invest $200 billion to extend the total length of HSR from 7,000 to 16,000 km. Japan has

1

Figure 1.1   High-speed trains in Japan.

Figure 1.2   High-speed trains in France.

Figure 1.3   High-speed trains in Germany.

Figure 1.4   China's high-speed trains.

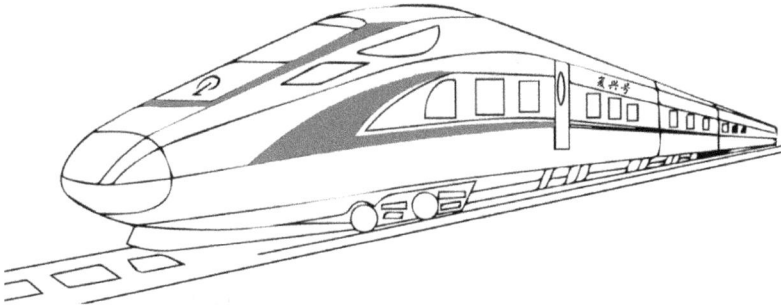

Figure 1.5 Wheel rail HSR.

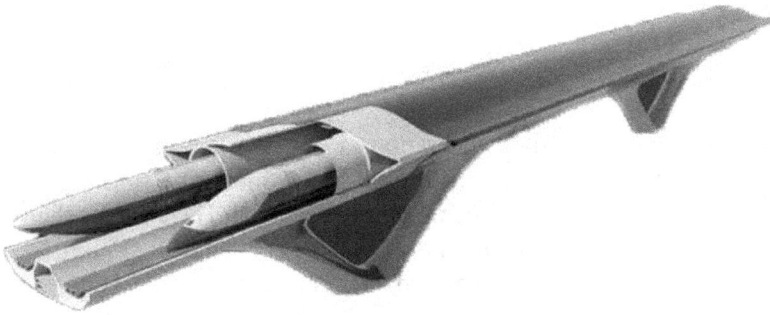

Figure 1.6 SSR system.

started construction of the Maglev Central Shinkansen, an HSR linking Tokyo and Osaka that will run at 550 km/h, making it the world's fastest train. In addition, China's HSR has been developed from scratch, going from seekers to front-runners, leading the world in both scale and speed. By 2024, China's HSR operating mileage will reach 38,000 km, enabling 90% of Chinese people to travel by HSR. Now, the world has entered the "HSR era" (Figure 1.5). And which mode of transportation will dominate the future? The future may belong to the world of the Super-Speed Rail (SSR).

SSR is a kind of vehicle designed on the theory of "Evacuated Tube Transport" (Figure 1.6), which has the characteristics of ultra-high speed, high safety, low energy consumption, low noise, no vibration, no pollution, etc.

People's pursuit of high-speed vehicles has never stopped. How to create faster, safer, more energy-saving, more comfortable, and more environmentally friendly vehicles has become the goal of mankind. SSR is a vehicle design based on the theory of "Evacuated Tube Transport," and it

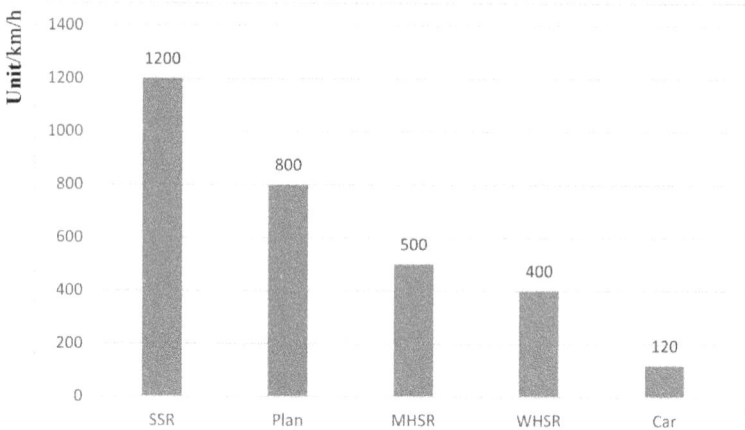

.**Figure 1.7**   Running speed of different vehicles.

may also be a new or fifth generation of transportation after cars, ships, trains, and aircraft. Since the fifth type of transportation runs faster than the conventional HSR (the operational speed of an HSR is less than 500 km/h, which is called conventional HSR), it is called SSR (Figure 1.7).

SSR operates in a fully enclosed vacuum system, which is immune to complex weather conditions. It uses solar energy as the driving force and has the characteristics of ultra-high speed, low energy consumption, low noise, low pollution, safety, and environmental protection. Over the years, scholars at home and abroad have conducted relevant studies on the concepts of SSR. For example, in 1904, the American scholar Robert David put forward the idea of "Evacuated Tube Transport," which was the first time for mankind to consider traffic operation without air resistance. In 1922, German engineer Herman Kener put forward the concept of the maglev train. He believed that the target speed of the maglev train of above 1,000 km/h could be achieved in a vacuum pipeline, and long-distance operation could be realized. In 1985, Daryl Oster, an American engineer, studied the feasibility of "Evacuated Tube Transport." He believed that the vacuum pipeline can carry out point-to-point transmission to achieve the purpose of direct transportation. In 1992, American scholar Helen David believed that in a vacuum state, the speed of running vehicles could reach 6,500 km/h, or even 20,000 km/h, and there was no limit to the speed without air resistance. In 2004, many scholars from Southwest Jiaotong University of China studied the system dynamics of vacuum pipes and improved the dynamic problem of vacuum pipes theoretically. In 2013, Elon Musk, CEO of Tesla (American Electric Vehicle Company), enriched the concept of "vacuum transportation" and put forward the concept of "Hyperloop." He proposed that the expected speed of the "Hyperloop" could be 1,200 km/h, close to the sound speed. Because the speed of 1,200 km/h will be two or three times faster than the running speed of the fastest bullet train (300–400 km/h) and two times faster than the flight speed of an aircraft (600–800 km/h), once the SSR is successfully developed, it may become the fastest means of transportation for mankind at present. The SSR system brings breakthroughs to human development. A global village has been realized in the SSR environment. SSR technology represents the commanding heights of the world's cutting-edge technology, thereby ensuring its leading position in the world.

Grasp the commanding heights of the world's cutting-edge technology, and then ensure the world leader's position. The core concept of SSR is high-speed driving. In the design of the SSR system, the super train will use a floating cabin in a low-pressure pipe to transport passengers at a speed of more than 1,200 km/h, to achieve the purpose of speed and efficiency. Is it possible to realize the SSR? Wang Mengshu, an academician at the Chinese Academy of Engineering, raised two questions: First, how can one solve the "vacuum breakdown" phenomenon of voltage? Second, how can one ensure human atmospheric environment inside the train? Shen Zhiyun, an academician at the Chinese Academy of Sciences, believed that there should be no major technical problems, but the key is economic feasibility. Of course, if these problems are solved and the lines are built, it seems to be a test whether people would dare use it.

1.1 System Research of SSR

The SSR system is a vehicle designed on the theory of "Evacuated Tube Transport," which connects a series of "vacuum pipelines" to form the whole transportation line system, allowing passengers to get from point A to point B in a matter of minutes, as shown in Figure 1.8.

As for the transportation system, on the one hand, the SSR system facilitates passengers' travel, saves travel time, and improves transportation efficiency relating to the means of transportation. On the other hand, the SSR system is a super train running in a vacuum pipeline, based on the concept of the evacuated tube (Figure 1.9). It is not affected by air

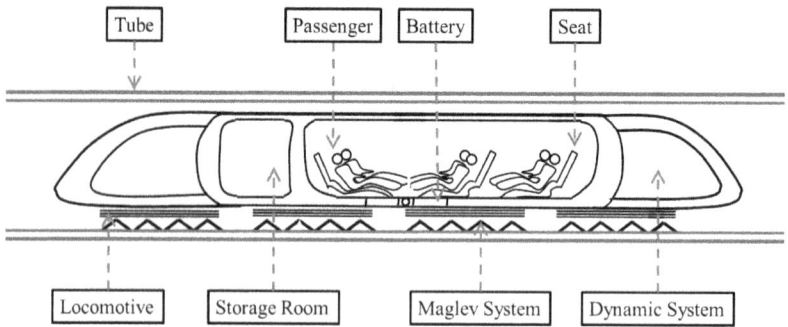

Figure 1.8   SSR system architecture diagram.

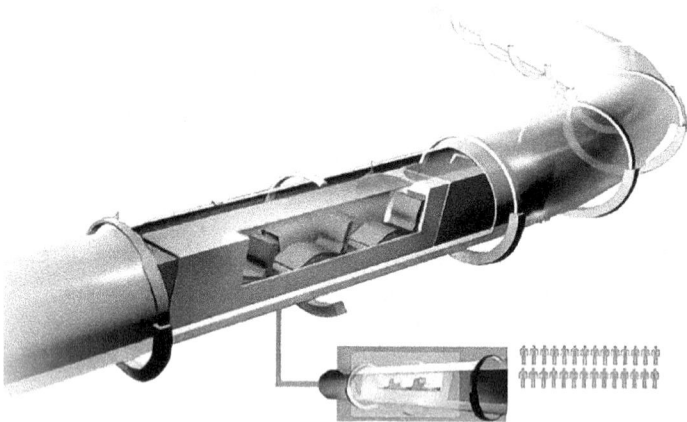

Figure 1.9   Concept diagram of SSR system.

resistance, friction, and weather (such as strong wind, rainstorm, debris flow, and low temperature). The speed of the super train can reach 1,000–20,000 km/h, which is several times higher than the flight speed of aircraft. It is an ideal mode of transportation. How can the SSR system reach or even exceed the speed of the aircraft? There are two ways. The first is to make the SSR system run at high altitudes like an aircraft, but that hard to achieve. The second is to simulate a vacuum environment for the SSR system to run in it. Therefore, the SSR system needs to build a pipe to exhaust the air inside, so there is no air resistance. If there is a vacuum in the SSR system, the theoretical speed can reach 6,500 km/h. Of course, it is too difficult to do that. If the SSR system simulates 0.1 atm, it can at least reach the speed of 1,000 km/h to 2,000 km/h.

1.1.1 *Coupling relationship between velocity and air*

In the dense surface atmosphere, high-speed vehicles will be affected by friction (contact friction and air friction, mainly air friction) during operation. The maximum speed of surface vehicles is about 500 km/h, while the theoretical maximum speed in a pipeline transportation system can reach more than 20,000 km/h. For example, when the speed of the HSR exceeds 300 km/h, the main resistance comes from the air, when the speed reaches 400 km/h, the resistance from the air exceeds 90%, and when the speed reaches 500 km/h, the resistance from the air exceeds 99%. The running speed of the HSR cannot exceed the flight speed of an aircraft, mainly because the surface air resistance encountered by conventional HSR is much greater than that encountered by an aircraft in the air. Therefore, the running speed of the vehicle is related to the air resistance. The greater the air resistance, the lower the running speed of the vehicle as shown in Figure 1.10, it shows the situation of air resistance when the vehicle is running.

(1) **Coupling relationship between surface velocity and air resistance:** When the vehicle runs on the ground, it faces an atmospheric pressure. Compared with cars, ships, ordinary trains, and other vehicles, the conventional HSR is the king of the speed on the ground. The operating speed limit is 400 km/h. Of course, it may also run at 486.1 km/h (the highest speed of China's conventional HSR on October 3, 2010), 574.8 km/h (the highest speed of France's conventional HSR on April 3, 2007), and 603 km/h (the highest speed of the maglev train in Japan on April 21, 2015), but these speeds are no longer normal economic speeds, but experimental speeds. According to the latest research results, the maximum economic speed of ground conventional HSR (wheel-rail type) in normal operation is 400 km/h, while the maximum economic speed of ground conventional HSR (maglev type) in normal operation is 500 km/h. Therefore, due to air resistance on the surface, the fastest economic speed of any vehicle cannot exceed 500 km/h. The optimal operating speed at different distances is shown in Table 1.1.

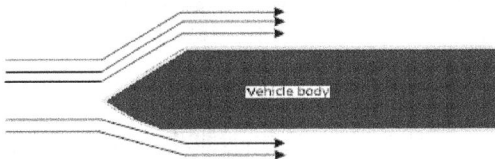

Figure 1.10   Relationship between velocity and resistance.

Table 1.1 Optimal velocity values at different distances.

No.	Distance (km)	Operating speed (km/h)	Vehicle
1	<200	200	Conventional HSR (EMU)
2	200–400	400	Conventional HSR (wheel-rail)
2	400–600	500	Conventional HSR (Maglev)
3	600–1,500	1,200	SSR
4	1,500–10,000	2,000	
5	15,000–20,000	6,500	
6	>10,000	20,000	

Table 1.2 Optimal velocity values at different atmospheric pressures.

No.	Altitude (m)	Atmospheric pressure (Pa)	Flight speed (km/h)	SSR operating speed (km/h)
1	<1,000	1	400–500	<500
2	1,000–4,000	0.8–1	500–600	500–1,200
2	4,000–10,000	0.5–0.8	600–800	
3	10,000–12,000	0.2–0.5	800–1,000	1,200–2,000
4	12,000–15,000	0.05–0.2	1,000–2,000	
5	15,000–20,000	0–0.05	2,000–10,000	2,000–20,000
6	>20,000	0	>10,000	>20,000

(2) **Coupling relationship between air velocity and air resistance:**
Because the air density and air resistance vary at different heights, and this air resistance is also related to atmospheric pressure (also known as atmospheric pressure), the operating speed of vehicles under different atmospheric pressures is also different. According to the results of the existing research, the optimal speed values under different atmospheric pressures are shown in Table 1.2.

Through Tables 1.1 and 1.2, the air resistance on the surface and in the air is analyzed: The thinner the air, the smaller the air resistance, and the greater the operating speed of the vehicle. Therefore, if a pipeline is built to exhaust the air inside, the pipeline is a vacuum, so that there is no air resistance in the operation of the vehicle and its speed can reach more than 6,500 km/h. Even if there is a little air in the pipeline, the vehicle can reach the speed of 1,000 km/h to 2,000 km/h as long as the pressure is less than 0.1 atm.

1.1.2 *Relevant theories in SSR system*

In the dense atmosphere on the surface, the operation of vehicles is affected by contact friction and air friction, and the main limiting factor is air friction, that is, air resistance. How can one improve speed? Only by reducing friction and resistance. On the one hand, for contact friction, the SSR system essentially uses the thrust provided by magnets and relies on compressed air to provide lift. The SSR system will not have friction resistance between "wheels and tracks." On the other hand, to achieve the target speed per hour, the SSR system should maintain low pressure in the running track to reduce the resistance between the super train and the air, as shown in Figure 1.11, it shows the situation of resistance when the super train is running.

(1) **Design principle — Air resistance problem:** The pipeline of the SSR system can be built as a closed pipeline. The air in the pipe can be removed, turning the operating pipe into a vacuum or partial vacuum.

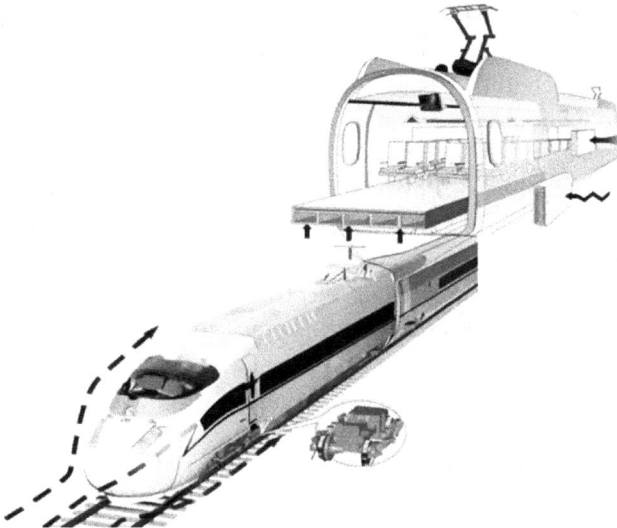

Figure 1.11 Resistance of train operation.

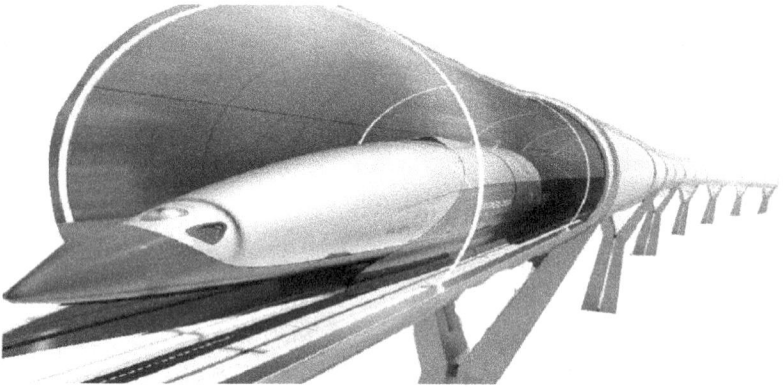

Figure 1.12   Operating system in vacuum.

In this way, when the SSR operates in a closed pipeline without air resistance, the resistance during the operation of the SSR, the energy consumption, the aerodynamic noise, and super train vibration will all be greatly reduced, as shown in Figure 1.12.

(2) **Design principle — Contact friction problem:** The friction resistance of the SSR comes from air friction and contact friction. In addition to eliminating the resistance caused by air friction, another highlight of the SSR is levitation technology. The levitation technology is used to solve the resistance of contact friction. The magnetic levitation technology is used to make the vehicle run in the vacuum pipeline without contact and friction, to achieve point-to-point transport, as shown in Figure 1.13.

(3) **Design principle — Power actuation problem:** SSR can adopt the self-powered design. According to the research of American expert Elon Musk, laying solar panels on the upper part of the operation pipeline can generate enough electric energy to maintain its normal operation. After the SSR system is equipped with solar panels on the operation pipeline, the energy obtained will satisfy the energy consumption of the whole system. In addition, energy storage facilities are added to the SSR system to store the excess energy for emergency applications, as shown in Figure 1.14.

Therefore, the construction of the SSR system is very simple in theory. First, it simply takes the air out of the closed environment to create a vacuum. Second, it eliminates friction and allows the vehicle to float in

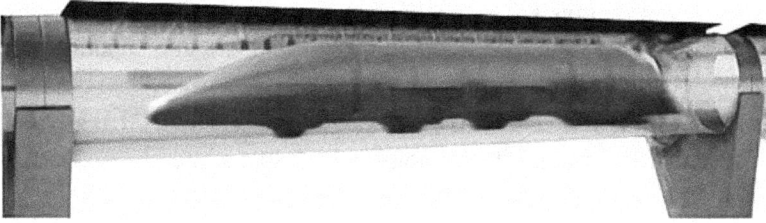

Figure 1.13   Zero friction of operation system.

Figure 1.14   Solar pipeline system of SSR.

the pipeline, so that the vehicle can move forward at a high speed with very little energy. Finally, driven by solar energy, the SSR runs rapidly in the vacuum pipeline.

1.1.3 *Definition of SSR*

The SSR system needs a pipe isolated from the external air, with vacuum in the pipe, and a maglev train and other vehicles in it (the schematic diagram of SSR, Figure 1.8, based on the idea of Elon Musk). The vehicle (i.e., super train) is in an environment with almost no friction, and a floating cabin in the low-pressure pipe is used to transport passengers at a speed of 1,200 km/h. The SSR has some characteristics of the five existing transportation modes (rail, aviation, water transportation, road, and pipeline):

Feature 1. Characteristics of SSR system based on pipeline transportation: The SSR system is involves rapid transport in a pipeline, which has some characteristics of pipeline transportation.

Feature 2. Characteristics of SSR system based on rail transit: The SSR system uses maglev technology, which has some characteristics of rail transit.

Feature 3. Characteristics of SSR system based on road traffic: The transportation capacity of the SSR system is equivalent to that of the bus, which has some characteristics of road traffic.

Feature 4. Characteristics of SSR system based on air traffic: The operation speed of the SSR system is similar to that of an aircraft, and it has some characteristics of air traffic.

Feature 5. Characteristics of SSR system based on water transportation: The super train floats in the air and has some characteristics of water transportation in principle.

Therefore, the SSR system combines the characteristics of the existing five modes of transportation into a new, possibly sixth, mode of transportation, as shown in Figure 1.15, the operation diagram of SSR system is introduced.

(1) **Super train:** SSR is a new means of transportation built using the concept of "Evacuated Tube Transport." The vehicle is a new generation of transportation after automobiles, ships, trains, and aircraft. It has the characteristics of ultra-high speed, high safety, low energy consumption, no noise, and no pollution. As it operates in a vacuum pipeline and adopts magnetic levitation technology, it is recommended that the vehicle be called a vacuum flying car or super train. The schematic diagram of a super train is shown in Figure 1.16.

(2) **Vacuum pipe:** Different from the traditional railway, SSR is a vacuum suspension frictionless flight system, which is a new high-speed transportation system. The SSR system is composed of a transportation pipeline, manned cabin, vacuum equipment, suspension components, an ejection system, and a braking system, as shown in Figure 1.17.

The SSR system has the following features: On the one hand, in the pipeline, the super train is suspended in the air, and the speed can reach more than 1,000 km/h. On the other hand, through maglev technology, the

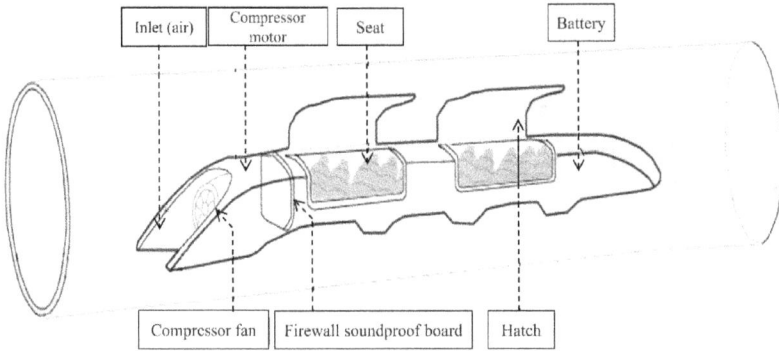

Figure 1.15   Operation diagram of SSR system.

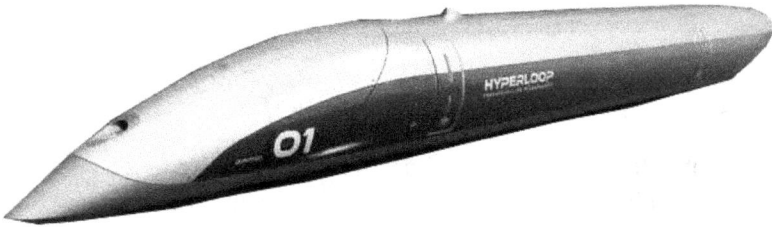

Figure 1.16   Schematic diagram of super train.

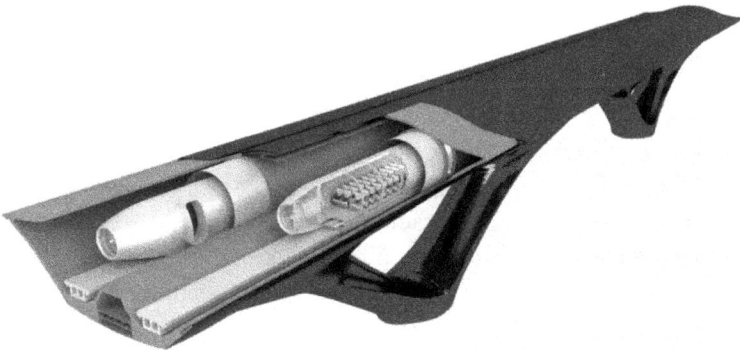

Figure 1.17   Schematic diagram of vacuum pipeline operation of SSR.

super train floats in the vacuum-treated pipeline, using the ejection device to launch to the destination, moving continuously along the pipeline. The SSR system architecture is shown in Figure 1.18.

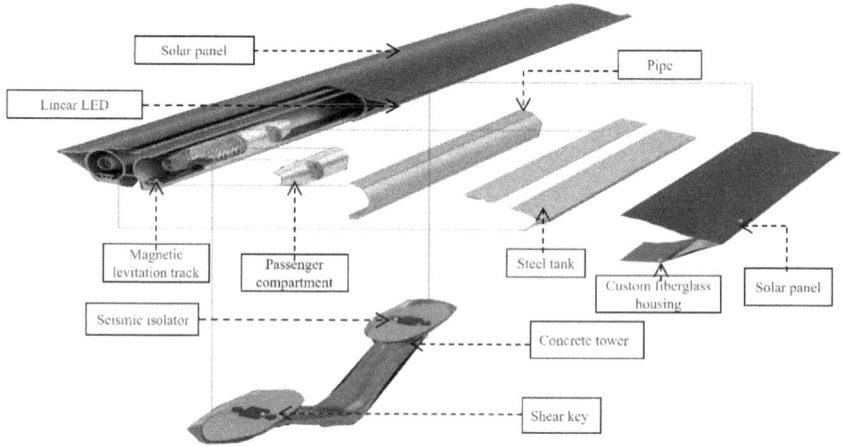

Figure 1.18   SSR system architecture.

1.2 Technical Characteristics of SSR

Since the maglev train is the fastest means of transportation in the world, the super train in the SSR system is the maglev train in a vacuum. In addition to the advantages of speed, the vacuum pipeline system also has the characteristics of punctuality, large transportation volume, comfort and safety, all-weather operation, energy conservation, and environmental protection.

1.2.1 *Safety of SSR*

No matter what kind of transportation one uses, safety is the most important facet. Without life, there is nothing else to discuss. Therefore, safety and reliability are the primary factors for passengers to travel. The SSR system uses huge, almost vacuum pipelines to connect multiple cities to form the SSR network, which is convenient to travel quickly, as shown in Figure 1.19. But what about the safety of the SSR? The following points relate to the safety of the SSR system:

(1) **Objectivity of SSR system safety:** From an objective point of view, the natural environment has a great impact on various means of transportation, but the SSR operates in a fully closed system and is not affected by the natural environment. The objective safety of the SSR system is mainly reflected in the following:

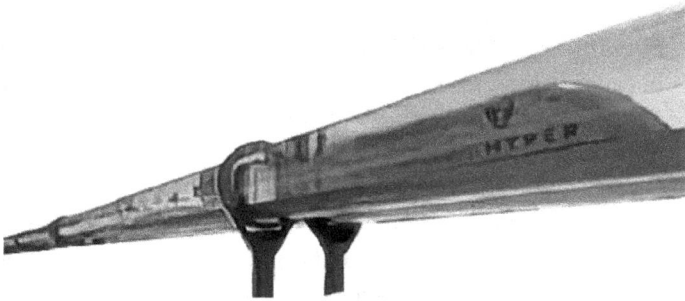

Figure 1.19 SSR system in pipeline.

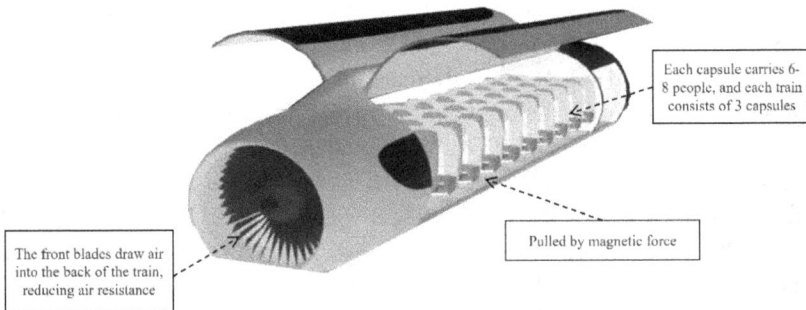

Each capsule carries 6-8 people, and each train consists of 3 capsules

The front blades draw air into the back of the train, reducing air resistance

Pulled by magnetic force

Figure 1.20 Internal safety of SSR system.

(i) *Safety in the natural environment*: Compared with aircraft, trains, and cars, the SSR lines are less affected by earthquakes and are not prone to accidents, because they do not fall or derail.

(ii) *Safety in the artificial environment*: Compared with air travel, SSR is not affected by weather factors (such as wind, ice, snow, fog, rainfall, and another natural climate), and situations such as delays and cancellations will not occur. Therefore, objectively, the SSR system has the highest safety.

(2) **Subjectivity of SSR system safety:** From the subjective point of view, most traffic accidents are related to people, while SSR is mainly intelligent control and has little to do with people, as shown in Figure 1.20. The subjective safety of the SSR system is mainly reflected in the following:

(i) *Safety based on design concept*: In terms of design, each part of the SSR has a speed limit, and the passenger compartment is completely contained in the pipeline, so it is unlikely to derail like a train.

(ii) *Safety based on rapid operation*: Unlike traditional trains and aircraft, the SSR will not be affected by human accidents, because it is a closed system and the advanced control and guarantee system makes provides it with safety unmatched by other transportation modes and tools.

(iii) *Safety based on system management*: Safety cabins are also set at certain distances along the vacuum pipeline. When the super train stops due to failure or the pressure of the sealed cabin is lost, passengers can escape to the safety cabin to avoid danger. Therefore, subjectively, SSR is the safest mode of transport.

In conclusion, whether subjectively or objectively, the SSR system is the safest means of transportation. Compared with vehicles, aircraft, ships, conventional HSR, and other means of transportation, the SSR is very safe. For example, the conventional HSR has had six traffic accidents since it began operation in 1964, as shown in Table 1.3. This is in sharp contrast to aviation with frequent air disasters (IATA's 2015 Global Air Safety Statistics report shows that there were at least 10 crashes worldwide in 2015 that killed more than 10 people, a total of 576 deaths). Road accidents have long developed into the first killer of mankind (according to the 2015 Global Road Safety Status Report published by the World Health Organization in 2015, about 1.3 million people are killed in road traffic accidents every year, and 20 million to 50 million people suffer non-fatal injuries). Moreover, the SSR operating in a pipeline is much safer than conventional HSR. Therefore, the SSR is the safest vehicle among the existing vehicles.

Table 1.3   High-speed railway accidents.

Name	Date	Country	Number of deaths	Number of injured	Cause
The first HSR accident	1998.06.03	Germany	101	194	Wheel rail
The second HSR accident	2005.04.25	Japan	107	549	Artificial
The third HSR accident	2011.07.23	China	42	192	Lightning
The fourth HSR accident	2013.07.24	Spain	79	180	Artificial
The fifth HSR accident	2015.11.14	France	42	32	Derailment
The sixth HSR accident	2016.07.14	Italy	27	50	Derailment and artificial

1.2.2 *Comfort of SSR*

Comfort is another key reason why we choose SSR. Many people think that SSR is faster than aircraft and the human body cannot bear it, but in fact the human body can bear it as proved through scientific analysis. This is because the limit that the human body can bear is an acceleration of about 50 m/s^2, while the 100 km acceleration of the car is about 10 s. If you can easily accelerate to 1,000 km/h within 1 min to 2 min, it is not a difficulty for the human body to bear the running speed of SSR. Moreover, for passengers, although the speed is great, they will not feel the high-intensity acceleration and noise in the vacuum environment, as shown in Figure 1.21. The comfort of the SSR is also reflected in the following:

Good physical adaptability: Each passenger compartment of the super train is pressurized and equipped with an oxygen mask and emergency braking system, so there will be no physical discomfort.

Good physical stability: The super train is ejected from the starting point. Because of the magnetic force, it will not encounter turbulence like an aircraft on the way.

Good physical sensitivity: When the super train starts, passengers will feel the acceleration. Once the super train moves forward at full speed, people will no longer feel it.

Therefore, in the SSR system, the passenger experience is very comfortable. The SSR will be more comfortable and quiet than the conventional HSR and aircraft. Especially from the technical level, the comfort of SSR also includes factors such as vibration, temperature, noise, air, and light, as shown in Table 1.4 for a comparison of various means of transportation.

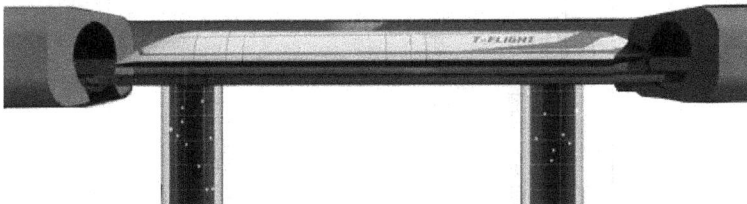

Figure 1.21   Piping system of SSR.

Table 1.4 Comparison of comfort of various means of transport.

Vehicle		SSR	Ordinary railway	HSR	Automobile	Aviation
Interior stability	Longitudinal stability	1.5	3	2	3	3.2
	Lateral stability	0.2	2.2	2	2.5	2.6
	Vertical stability	1	2.5	2	2.8	5
Interior noise		50 dB	70 dB	65 dB	76 dB	80 dB
Interior temperature		Self-adjusting temperature	Higher than normal temperature	Self-adjusting temperature	Higher than normal temperature	Self-regulating normal temperature
Interior air		Second to outdoor	Same as outdoor	Second to outdoor	Same as outdoor	Second to outdoor
Interior light		Self-adjusting light	Same as outdoor	Second to outdoor	Same as outdoor	Self-adjusting light
Remarks		The smaller the stability value, the more stable and comfortable the interior environment. The internationally recognized stability threshold is 2.				

1.2.3 *Economic value of SSR*

The economic value is also one of the main conditions to be considered in the construction of the SSR system. According to the available data, the cost of evacuated tube transport will be very cheap, only 1/4 of the cost per kilometer of an expressway and 1/10 of the cost per kilometer of the conventional HSR. Moreover, the operation cost of the SSR system is lower than that of conventional HSR. Especially if the distance between the two cities is longer, the operation cost of SSR is lower, which is 1/2 of the ticket price of conventional HSR. According to the design concept of Elon Musk, the SSR system is very economical for any two large cities with a distance of no more than 1,500 km. For example, the cost of building the SSR system between Beijing and Shanghai is 6 billion yuan. If the SSR runs every 3 min, each SSR carries 30 people, and the operation cost of each trip is about 200 yuan. Therefore, the one-way fare can be set at 200 yuan per trip, which is very cheap and acceptable to passengers.

(1) **Economic analysis based on construction cost:** The operation line of the SSR system is the pipeline, which is supported by elevated pillars, away from the ground, to reduce the occupation of land resources. The vacuum pipeline between the two cities is built on the ground like the HSR, and as long as there is a road, a two-way pipeline can be built on that. Moreover, the vacuum pipeline may also be "attached" to the built high-speed bridge to save resources and infrastructure construction costs. Therefore, the construction cost of SSR is lower than that of other means of transportation.

(2) **Economic analysis based on operating cost:** The super train uses solar energy to greatly reduce its transportation cost and can use its technology to store energy many times. After the SSR system accelerates the super train to a certain speed, the super train can run in the vacuum pipe relying on inertia without any additional energy. When the super train is about to arrive at a station and needs to slow down, the existing kinetic energy of the super train can be recovered and reused through the motor, so the transportation cost of the super train is only 1/10 of that of the conventional HSR. Therefore, the operation cost of SSR is lower than that of conventional HSR.

1.2.4 *Convenience of SSR*

Convenience is one of the conditions for choosing SSR. If the SSR network can be built all over the world, global travel can be completed in a few hours and one can realize global one-day travel (going out early and returning late, and going to work all over the world). According to Elon Musk's design concept, the convenience of SSR is mainly reflected in the following:

The convenience of the SSR system I: "Come and go, everyone is equal": The SSR system does not require one to book seats. There is no class difference in seats on SSR. First come, first served, everyone is equal.

The convenience of SSR system II: "Save time and no cost": For long-distance transportation, the SSR system has more advantages than aircraft, because SSR does not waste time on take-off and landing. The SSR station is located in the city center, so people can reach the station without transfer, which also saves the cost of going to the airport.

The convenience of SSR system III: "Willful travel and free choice": The SSR system operates automatically, and passengers do not need to worry about delays and other problems. Passengers taking the super train do not need to take it according to time like taking a plane, but can choose to travel at will.

Therefore, no matter compared with any kind of transportation, the SSR system is convenient. In particular, in order to improve the operation efficiency of the SSR system, it can operate at different speeds according to different operation distances. The recommended values are shown in Table 1.5.

Table 1.5   Operation speed of SSR for different travel distances.

	Short distance (km)	General distance (km)	Medium distance (km)	Long distance (km)
Name	<500	500–1,000	1,000–10,000	>10,000
Operating speed (km/h)	300–500	600–800	5,000–6,500	6,500–10,000

1.2.5 *Energy saving and environmental protection feature of SSR*

In addition to the advantages of speed, the SSR system is more energy saving and environmentally friendly. In particular, SSR has obvious advantages in low carbon emission, energy conservation, and environmental protection. On the one hand, as a means of transport, super trains not only have zero carbon emissions but also have no pollutants such as dust, oil fumes, and other exhaust gases. On the other hand, evacuated tube transport is an air-free and friction-free transportation mode, which is quieter and noiseless compared to conventional HSR and aircraft.

(1) **Technical characteristics analysis of SSR system based on energy consumption:** Because the SSR system reduces contact friction and air friction, evacuated tube transport consumes less energy than any traditional means of transportation. The transportation capacity per kilowatt-hour of the super train is 50 times that of conventional HSR. The SSR system will be powered by solar energy, which can supplement itself, and the system also has energy storage facilities, which can drive for a week without using battery panels. According to analysis based on existing research results, a comparison of energy consumption of various means of transportation is shown in Table 1.6.

(2) **Technical characteristics analysis of SSR system based on environmental protection:** Super trains fly more than twice as fast as aircraft, but their energy consumption is less than 1/10 of that of civil airliners, and the noise and exhaust pollution and accident rate are close to zero. In particular, the SSR pipeline is built underground or above the ground, which does not pollute the environment. According to analysis based on existing research results, a comparison of environmental protection features of various means of transportation is shown in Table 1.7.

Table 1.6 Comparison of energy consumption of various vehicles.

Vehicle	Ordinary railway	HSR	SSR	Automobile	Aviation
Energy consumption per person for the same mileage (kg/person)	1	0.5	0.1	6	4

Table 1.7 Comparison of environmental protection features of various means of transport.

Vehicle	Ordinary railway	HSR	SSR	Automobile	Aviation
CO_2 emission per person km [mg/(km·person)]	1	0.5	0.2	10	4
Noise per person km [db/(km·person)]	0.1	0.05	0.01	1	1

1.3 Problems of SSR

Once the SSR system is successful, it will completely challenge human cognition of traffic, but the greater the speed, the higher the risk coefficient. If an accident occurs, the vacuum pipeline will bring unimaginable disaster to the passengers. At present, from a technical level, the various key technologies used in the SSR system (including a low-pressure pipeline, compressor, and solar energy) are mature and feasible. However, from an application level, there are many other problems to be solved. Academician Shen Zhiyun believes that there should be no major technical problems, but the key is economic feasibility. However, science and technology are constantly improving. Soon, the idea of SSR will no longer be a fantasy, but a means of transportation that can be put into experiment. However, the following problems need to be solved:

(1) **Air resistance problem:** The speed of SSR must reach more than 1,000 km/h to have a practical and easy operation. However, at present, the fastest vehicle running on the ground is the superconducting maglev "bullet" train in Japan, with a maximum speed of 603 km/h. Jim Powell, one of the inventors of the "bullet," believes that this speed (603 km/h) is the extreme limit of the open rail train, because once the speed exceeds 300 miles (about 483 km/h), the problem of air resistance will become very serious and cause significant noise (Figure 1.22). Therefore, reducing air resistance is the main condition to realize the rapid operation of SSR. Therefore, the problem of air resistance needs to be further studied.

(2) **Operation route problem:** The straighter the route of the SSR system, the better, because turning under high speed is not only very difficult but also very uncomfortable for passengers (Figure 1.23). In his research, Powell found that passengers can bear up to one-tenth of the lateral force

Figure 1.22   Resistance to train operation.

Figure 1.23   HSR moving forward in a straight line.

of gravity when turning; that is, the turning radius of the SSR system when running at 750 miles (about 1,200 km/h) would reach at least 50 miles (about 80 km). Therefore, the SSR can only go in a straight line. As the SSR will inevitably pass through some difficult terrain, it will bring many difficulties to the construction of the SSR system.

1.4 Research and Development Prediction of SSR

Transportation represents the most fundamental dream of mankind: to break through the constraints of space and time and reach farther places at the fastest speed. As the fastest means of transportation, SSR has many advantages, which not only improve transportation efficiency but also reduce environmental pollution and energy consumption. However, the current technical level and economic cost are not enough to put the SSR into use and promote its application.

(1) **Theoretically, the SSR system is completely feasible:** In theory, pipeline transportation is the most efficient and energy-saving transportation mode at present. The vacuum maglev train is the fastest means of transportation in the world, which has been verified theoretically. SSR is a maglev train running in a vacuum pipeline. Therefore, from a theoretical point of view, it is completely feasible to build the SSR system because of the following:

  (i) *High-speed operation*: In the dense surface atmosphere, high-speed vehicles are more or less affected by friction during operation. Therefore, at present, the maximum speed of such vehicles is about 500 km/h, while the maximum speed of a pipeline transportation system can theoretically reach 20,000 km/h.

 (ii) *High system security*: The super train in a fully enclosed environment is completely unaffected by weather changes. Evacuated Tube Transport also has the advantages of safety and environmental protection.

(iii) *Reasonable energy supply*: In terms of energy, the SSR adopts a self-powered design, and solar panels are laid above the pipeline to generate enough electricity to maintain operations.

(2) **In terms of application, it is technically difficult to build the SSR system in a short time:** In terms of application, it is difficult to realize vacuum transportation above 1,000 km/h, especially in terms of technology, cost, and management. Therefore, from the perspective of the application, it is impossible to build the SSR system at present because of the following:

  (i) A long-distance vacuum pipeline is difficult to build. The pressure difference inside and outside the vacuum pipeline is great. It is difficult to ensure that there is no air leakage for thousands of kilometers. This is difficult to achieve with the current technical means.

 (ii) Magnetic levitation technology is not perfect. There are still many insurmountable obstacles in the actual operation of magnetic levitation technology, not to mention the transportation based on vacuum pipeline, which needs further research.

(iii) The voltage needs to be stable, but ensuring this stability is a big problem. Voltage easily produces a "vacuum breakdown" phenomenon and self-sustaining discharge in a vacuum environment, which may damage the electrode and lead to the paralysis of the SSR transportation system.

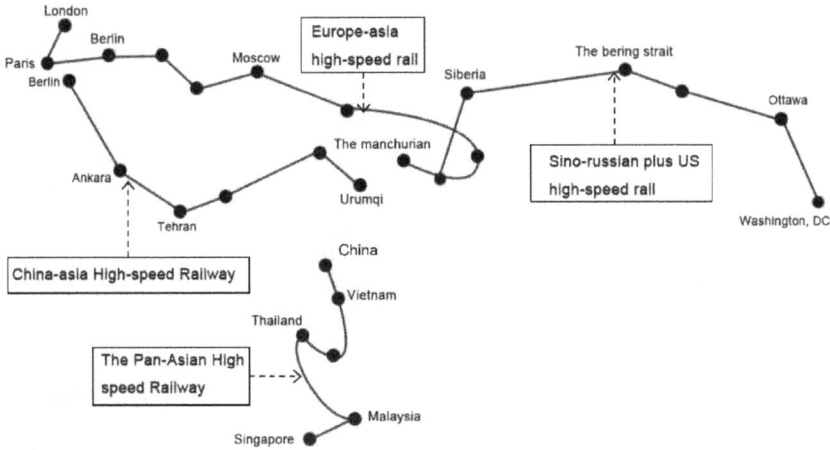

Figure 1.24   Global integration of SSR system.

(iv) System management issues will be encountered. The SSR system is a long-distance transportation mode, which may cross
different countries and regions. Managing these effectively is a
big problem.

In conclusion, the SSR system has many advantages, which may trigger a revolution in the field of transportation and promote the progress of
human society in the future. However, there are still many problems in
technology and cost, which need to be studied and discussed by scholars.
If the SSR is to develop, it must be gradual and cannot develop by leaps
and bounds, otherwise; people will pay a price. Only when the ordinary
railway gradually transits to the conventional HSR can it stimulate the
pace of moving toward the SSR, achieve the balance of "speed and
safety," realize global integration under the SSR environment, and achieve
the global one-day tour of "going out early and returning late," as shown
in Figure 1.24.

1.5 Summary

"The concept is feasible, but the theory is flawed." With the promotion of
conventional HSR, metropolitan areas of various countries have formed
rapidly, which has greatly shortened the distance between urban and rural
areas, and accelerated the integration of urban and rural areas. In the

future, with the promotion of the SSR system, a world economic circle will also be formed rapidly, which will greatly shorten the distance between countries, promote the rapid development of countries, and form a global village under the HSR environment. However, conventional HSR is an infrastructure project with a high cost and high investment, and SSR is a high-risk facility, which is directly related to a country's national economy and people's livelihood. Theoretically, SSR can reach a higher speed in a vacuum environment, but the maximum speed is not only related to the degree of vacuum but also related to suspension guidance system, traction system, track system, operation control system, and other technologies. Therefore, the SSR needs further theoretical research.

With the development of the HSR, a commercial operation speed of 350 km/h can be achieved. As an important means of transportation for human beings, the technical development direction of HSR in the future must be toward higher speed. The improvement of speed shortens time and space, shortens the distance, reconstructs the space–time map of the world, and realizes global integration "fancy ideal, but poor reality." SSR provides people with the imagination space to continue to improve the speed of travel, but there are still many problems to be solved to realize the engineering application, and mankind still needs to make continuous efforts to study, so that the dream becomes a reality.

Chapter 2

Basic Overview of Super-Speed Rail (SSR)

SSR is a vehicle designed with "Evacuated Tube Transport" as the theoretical core. It has the characteristics of ultra-high speed, low energy consumption, low noise, and low pollution. Because of its capsule appearance, it is called "Capsule HSR." This "capsule" train may be a new generation of transportation after cars, ships, trains, and aircraft. The maximum speed of a capsule HSR is expected to reach 4,000 km/h. If compared with the maximum speed per hour, the speed of the SSR is three times the speed of sound and four times the speed of a plane! In this way, it only takes 3 min from Beijing to Tianjin, 20 min from Beijing to Shanghai, and 30 min from Beijing to Wuhan. The **SSR system in operation** is shown in Figure 2.1.

The SSR system is also called evacuated tube transport. Evacuated tube transport has the characteristics of ultra-high speed, very low energy consumption, very low pollution, very low noise, and relative safety. The construction cost is not high. After completion, it can undertake most of the long-distance passenger and freight transportation tasks and reduce the proportion of long-distance buses, trains, and aircraft, and is expected to effectively solve the human traffic dilemma. Therefore, the research and development of SSR is a great project benefiting all of mankind.

2.1 Development History of SSR

With the development of the HSR, a commercial operation speed of 300 km/h can be achieved. As an important means of transportation for human

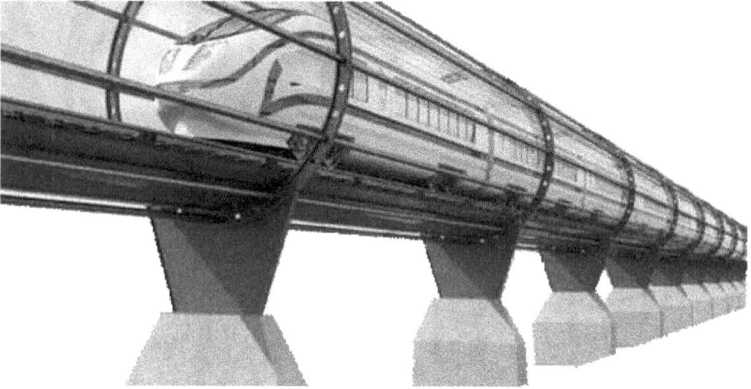

Figure 2.1   SSR system in operation.

beings, the technical development direction of HSR in the future must be toward high-speed trains with higher speed. In this way, scholars at home and abroad put forward the concept of "SSR" based on "vacuum transportation," subverting the traditional rail transit travel mode, which is more in line with today's fast-paced modern life.

2.1.1 *Past SSR design*

The SSR system is a vehicle designed with "Evacuated Tube Transport" as the theoretical core. The technology of "Evacuated Tube Transport" has to review the pipeline pneumatic passenger and freight trains that appeared in the early 19th century. At that time, steam locomotives had not been popularized, and electric power and internal combustion traction had not been invented yet Intelligent technology pioneers tried to drive the train using compressed air as power in the sealed pipeline, which was called "pipeline pneumatic transportation system." At this time, the sealed pipeline could not be made into a vacuum. Figure 2.2 shows the concept of SSR. Given in the following are the phases of development of the SSR:

Phase I: Theoretical stage of SSR: In the 1920s, the concept of "Evacuated Tube Transport" appeared. Robert Goddard, an American rocket expert, first put forward this idea. In the 1930s, German scientist Herman Kemper proposed that the maglev train system should be placed in a closed pipeline in a low-pressure environment, and it was envisaged

Figure 2.2 Conceptual diagram of SSR.

Figure 2.3 Schematic diagram of early pipeline transportation.

that the train speed would reach 1,800 km/h. In the 1950s, scientists at the Massachusetts Institute of technology also proposed a plan to build vacuum pipelines. In 1978, the technical experts of Rand Corporation proposed to build a transportation system called "transportation star," composed of underground pipelines with a low-pressure environment and a maglev train system. In 1999, Daryl Oster, a famous American engineer, obtained the invention patent of the Evacuated Tube Transport system. In 2010, Auster established a company dedicated to developing vacuum transportation projects. He envisaged that evacuated tube transport would be a transportation container similar to the capsule, which would carry out point-to-point transmission through a vacuum pipeline. Since the pipeline is in a vacuum, the speed of the capsule container can reach 6,500 km/h. Figure 2.3 shows a diagram of early pipe transportation.

China also developed maglev train technology with vacuum tubes. In December 2004, China held a seminar on "vacuum pipeline high-speed transportation" with the participation of eight academicians from the Chinese Academy of Sciences and Chinese Academy of Sciences and attended by several domestic authoritative experts, during which the feasibility of vacuum pipeline high-speed transportation was demonstrated. In 2005, academician Shen Zhiyun, a traction power expert, wrote an article on the technical scheme and realization of vacuum pipeline high-speed train. In 2011, Southwest Jiaotong University developed the "vacuum pipeline maglev vehicle experimental system," with the system pressure reaching 0.012 Standard atmospheres, which is the world's first complete vacuum pipeline test equipment combining vacuum pipeline, magnetic levitation, and linear drive technology at the same time. Figure 2.4 shows the SSR developed in China.

Phase II: Research and development stage of SSR: In 2013, Elon Musk, CEO of Tesla, an American electric vehicle company known as "Technology Maniac," enriched the concept of "vacuum transportation," put forward the concept of "Hyperloop" from SSR, and contributed more design details to this transportation concept. Elon Musk's speed expectation of SSR is more conservative than that of Auster. His expected speed is 1,200 km/h, close to the sound speed of 340 m/s. This speed will be two or three times faster than the fastest bullet train now and twice as fast as a plane. The high-speed train can take 28 people. The fare from Los Angeles to San Francisco will be $20. It can transport 7.4 million passengers a year and recover the investment in 20 years. A Schematic diagram of a super train is shown in Figure 2.5.

Elon Musk was born on June 28, 1971, in Pretoria, the administrative capital of South Africa (current name, Tshwane). He has dual citizenship of Canada and the United States, and is an entrepreneur, engineer, and

Figure 2.4   SSR developed in China.

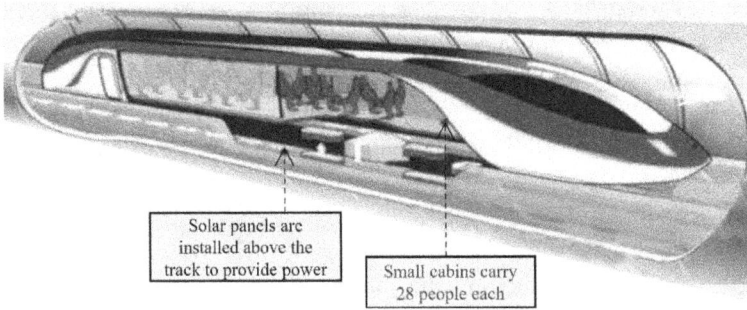

Solar panels are installed above the track to provide power

Small cabins carry 28 people each

Figure 2.5 Schematic diagram of super train.

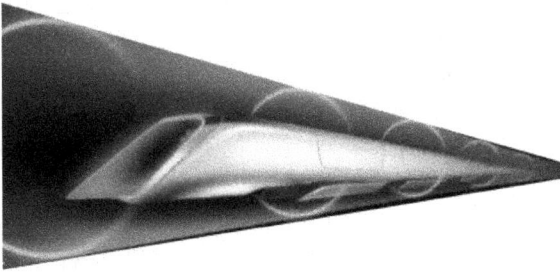

Figure 2.6 Simulation diagram of SSR system.

philanthropist. He is currently CEO and CTO of SpaceX, CEO of Tesla, and chairman of the board of directors of SolarCity.

Phase III: Test phase of SSR: The evacuated tube transport project of the ET3 Company in Colorado is also designed based on "SSR." According to the introduction of the ET3 Company, engineers will build a fixed vacuum pipeline similar to a railway track on the ground and place a "capsule" train in the pipeline. The "capsule HSR" train is shaped like a space capsule. It is estimated that the single weight is 183 kg, which is lighter than a car. It is about 4.87 m long and can accommodate four to five passengers. Figure 2.6 shows the SSR system.

2.1.2 *Present SSR design*

The SSR proposed by Elon Musk depicts a beautiful vision for human travel in the future. However, the development of SSR is facing countless difficulties and tests. A brief overview of the SSR development is different countries is provided in the following:

(1) **SSR in the United States:** On March 21, 2017, the US "Super Railway Transportation Technology Company" said that they had begun to build a passenger compartment for the "SSR." According to the scheduled plan, the "SSR" train can transport 164,000 passengers a day and leave every 40 s. This new traffic system can accelerate to 1,220 km/h, exceeding the maximum speed of most aircraft. On May 12, 2017, Hyperloop One of the United States conducted a comprehensive test of its SSR technology in a vacuum environment for the first time and achieved a speed of 113 km/h by using magnetic levitation technology. The test speed reached 310 km/h in July 2017. A simulation diagram of American SSR is shown in Figure 2.7.

(2) **SSR in China:** On August 29, 2017, China Aerospace Science and Industry Corporation (CASIC) announced that it had launched an R&D project for "high-speed flying train" with a speed of 1,000 km/h, and would develop super-high-speed trains with maximum operating speeds of 2,000 km/h and 4,000 km/h. At present, the world's first annular experimental line platform of vacuum pipeline ultra-high-speed maglev train has been built in China's Laboratory (Figure 1.8). The experimental line has a total length of 45 m, a design load of 300 kg, a maximum load of 1T, a net suspension height of more than 20 mm, and a theoretical train speed of more than 1,000 km/h under the ideal state of vacuum pipeline. Experts call this pipeline the prototype of SSR, as shown in Figure 2.8.

(3) **SSR in France:** In 2018, the construction of the first SSR test track in Europe began in Toulouse, France. Toulouse is not only the fourth largest city in France but also the headquarters of many transportation and aviation giants, including Airbus. Hyperloop Transportation Technologies (HTT) said that the test track of the SSR will be constructed in two stages: The closed 320 m system will be put into use in 2018. The 1-km-long full-scale system, built on an elevated

Figure 2.7   Simulation diagram of American SSR.

Figure 2.8   Simulation diagram of China's SSR.

Figure 2.9   Pipeline of French SSR system.

structure of up to 5.8 m, will be completed in 2019. On April 11, 2018, the SSR Transportation Technology Company began to deliver the first batch of rail pipes (Figure 2.9).

2.1.3 *Future SSR design*

In the era of HSR, every accelerated upgrade of transportation methods not only brings changes in travel time, but also promotes the vigorous development of the economy. The establishment of the SSR will be an opportunity of extraordinary strategic significance for all countries. Therefore, the United States, China, France, Russia, India, and other countries have launched SSR projects and are striving to occupy an absolute advantage in the speed competition in the future:

(1) **America's vision of SSR:** In February 2018, the Washington government stated that Elon Musk could build the "SSR" connecting New York and Washington (about 363 km). Hyperloop Transportation Technologies, an American SSR company, said that the maximum speed of the capsule SSR built in the United States for testing will reach 700 km/h, while the speed of the fully built SSR will reach 1,220 km/h. Magnetic levitation acceleration devices configured everywhere in the pipeline push the "capsule" forward. The interior of the pipe is evacuated to reduce air resistance. In addition, the compression fan device at the head of the train can suck in air and discharge it from the bottom of the train to form an air cushion several millimeters thick to suspend the train to reduce friction consumption. The "SSR" uses solar energy all the way, which greatly reduces the use of energy. The Vision of the American SSR System is shown in Figure 2.10.

(2) **India's vision of SSR:** When it comes to public transport in India, the biggest impression in people's minds is a train full of people (Figure 2.11). But India has decided to try to be at the frontier of rail transit. On July 31, 2019, the state of Maharashtra in western India finally approved the construction of a "Hyperloop" between Mumbai, India's financial center, and Pune 200 km away. After completion, it will take only 35 min from Mumbai to Pune, compared with 3.5 h in a car. As

Figure 2.10   Vision of American SSR System.

Figure 2.11 Indian train full of people.

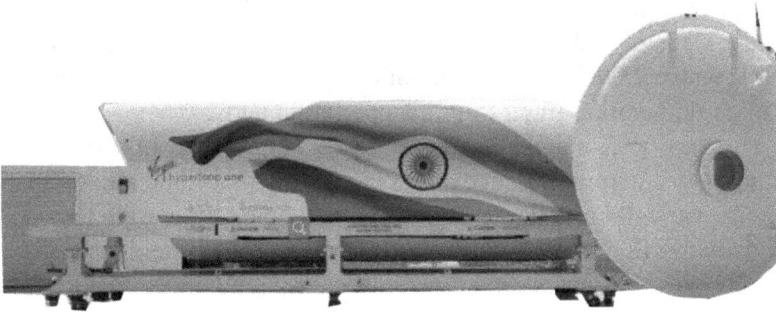

Figure 2.12 Design of SSR for India by Virgin Hyperloop One.

the first country in the world to support SSR technology, India identified SSR as a "public infrastructure" project, which is the same as the nature of roads, bridges, and railways, making it the first SSR project in the world. The local government has selected Virgin Hyperloop One (Figure 2.12) and its partner DP World as the initiator of this multi-billion-dollar infrastructure project.

(3) **China's vision of SSR:** China's "high-speed flying train" project, which is still in the research and demonstration stage, will be gradually realized according to the following three-step strategy:

Figure 2.13   Vision of China's SSR.

The first step, primary stage: build 1,000 km/h SSR. In the initial stage, the regional intercity flight train transportation network will be built mainly through the development of SSR with a transportation capacity of 1,000 km/h.

The second step, development stage: build 2,000 km/h SSR. This is also the stage of technology upgrading, mainly through the development of SSR with a 2,000 km/h transportation capacity to build a national super urban agglomeration flight train transportation network.

The third step, mature stage: build 4,000 km/h SSR. In this stage, the technology is mature. By developing the SSR with a 4,000 km/h transport capacity, we will build the "one belt and one road" flight train transportation network. Figure 2.13 shows the envisaged SSR system in China.

The launch of the SSR project will usher in an epoch-making transportation reform and will become the largest breakthrough in the field of railway and aviation in history! In the future, we believe there will be more subversive things in our life. It is undeniable that human life will be more and more convenient and efficient. Life never favors those who are stuck in the rut and content with the status quo. Instead, life favors those who are brave and innovative.

2.2 Related Concepts of SSR

The HSR system is a large system composed of a special line, high-speed train, and special control system, while SSR is a large system composed of a super train, super station, and super line. Therefore, like HSR, SSR is also a systematic concept. SSR is fast, cheap, and safe, and has open-source solutions.

2.2.1 *Super train*

Super train is the abbreviation of a super-high-speed train, also known as capsule train. It uses the concept of "Evacuated Tube Transport" to build a new means of transportation. Super train is an essential part of the SSR system. The design of the super train is meant to ensure a good journey for the passengers. Super trains have the advantages of comfort, safety, stability and reliability. The super train architecture is shown in Figure 2.14 with a description of the train and capsule in the following:

(1) **Super train:** The super train (Figure 2.15) is a vehicle with "Evacuated Tube Transport" as the theoretical core. Two American companies put forward the design model one after another. The super train is considered to be a new generation of transportation after cars, ships, trains, and aircraft. Taking the SSR is supposed to feel like taking a plane. The journey is smooth as if driving in the air. Elon Musk has described SSR as a cross-border transportation system "somewhere between a Concorde, a rail gun, and an air billiard."

(2) **Capsule train:** The concept of "capsule SSR" was originally proposed by Elon Musk, founder of PayPal, Tesla, and SpaceX, in 2012. Taking the "capsule train" of ET3 Company as an example, according to the designer, the transportation system is composed of a transportation pipeline, manned cabin, vacuum equipment, suspension parts, ejection system, and braking system. During operation, the capsule cabin weighing 183 kg, is 4.87 meters long and about 1.5 m high and capable of accommodating 4 to 6 passengers, is "floated" in the vacuum-treated pipeline through maglev technology. Then, the ejection device is used to launch the "capsule" and drive to the destination continuously along the pipeline. The conceptual diagram of SSR is shown in Figure 2.16.

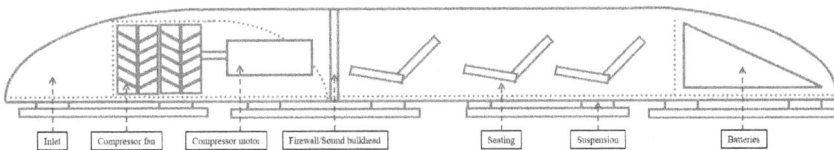

Figure 2.14   Super train architecture.

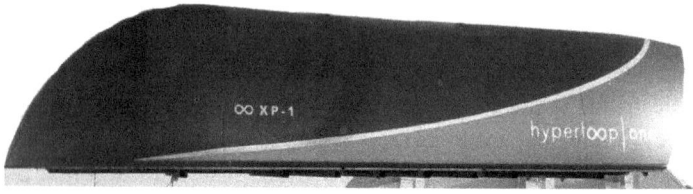

(a)

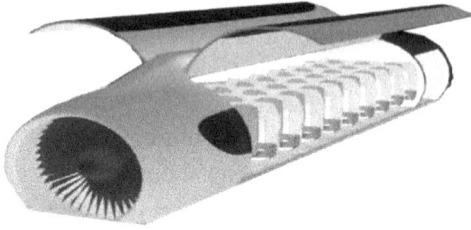

(b)

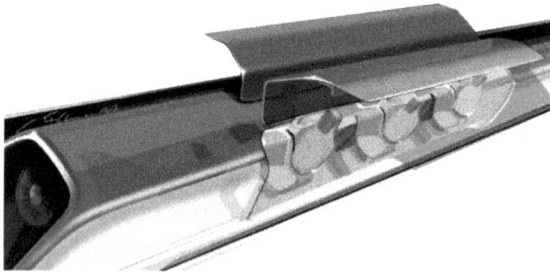

(c)

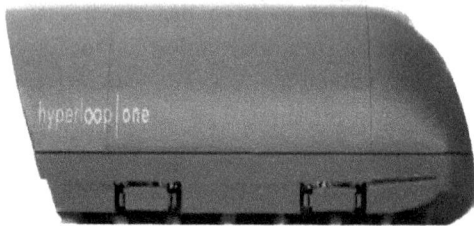

(d)

Figure 2.15   Super train simulation diagram series. (a) Super train simulation diagram based on "bullet." (b) Simulation diagram of super train based on "Dapeng winging." (c) Simulation diagram of super train based on "bird pours." (d) Simulation diagram of super train based on "passenger ship." (e) Simulation diagram of super train based on "cargo ship." (f) Simulation diagram of super train based on "mouse." (g) Simulation diagram of super train based on "elephant." (h) Super train simulation diagram based on "Battery."

(e)

(f)

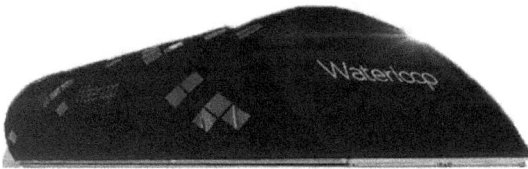

(g)

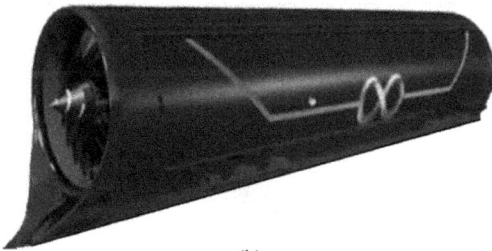

(h)

Figure 2.15 (*Continued*)

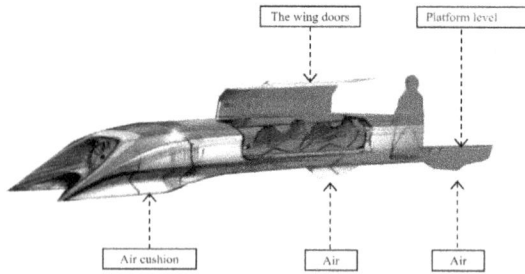

(a)

(b)

(c)

(d)

Figure 2.16   Capsule train simulation diagram series. (a) Super train simulation diagram based on "gull wing." (b) Super train simulation diagram based on "arrow." (c) Super train simulation diagram based on "bullet." (d) Super train simulation diagram based on "bullet." (e) Super train simulation diagram based on "bullet." (f) Super train simulation diagram based on "Capsule." (g) Super train simulation diagram based on "Capsule." (h) Super train simulation diagram based on "Capsule."

(e)

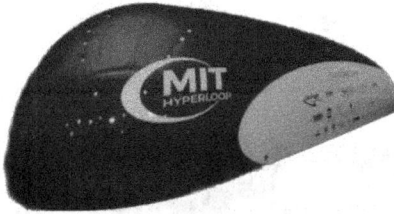

(f)

(g)

(h)

Figure 2.16   (*Continued*)

2.2.2 *Super station*

Super station is the abbreviation of SSR station. The design concept of the SSR station is very simple but practical. The ride process and layout are much simpler than boarding at the airport. According to the size of passenger flow, the stations of the SSR system can be divided into small stations and large stations.

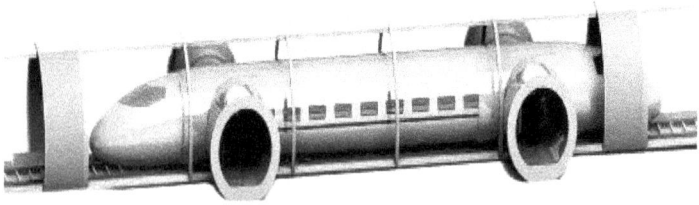

(a)

(b)

Figure 2.17 Conceptual design diagram of small station of SSR system. (a) Super station simulation diagram based on "forward and out." (b) Simulation diagram of super station vehicle based on "side up and down."

(1) **Small stations of SSR system:** In order to reduce costs and improve operation efficiency, SSR stations with small passenger flow are established called small stations. Figure 2.17 shows the conceptual design of small stations of the SSR system.

(2) **Large stations for the SSR system:** To improve transportation efficiency and promote the rapid operation of SSR, relatively large stations with large passenger flow are established, which are called large SSR stations. Figure 2.18 shows the conceptual design of large stations of the SSR system.

2.2.3 *Super line*

Super line is the abbreviation of SSR system operation line, which mainly refers to the vacuum pipeline for the rapid operation of super trains. Super

(a)

(b)

Figure 2.18 Conceptual design of large stations of SSR system. (a) Simulation diagram of super station based on "double side open screen." (b) Simulation diagram of super station based on "up and down."

trains run in vacuum pipes, and each super train is like a capsule. Each super train is placed in a pipe and fired to its destination like a shell. The super train runs continuously in an environment with little friction. The pipelines of these SSR systems constitute super lines. The laying of super lines not only needs to solve various problems inside the pipelines but also work through the challenging route planning. Simulation diagrams of super pipelines are shown in Figure 2.19.

Relevant studies at home and abroad show that super lines have the following requirements:

Requirements I: Design of "straight line forward" of super line: When the train turns, passengers can bear the lateral force of one-tenth of the

(a)

(b)

(c)

(d)

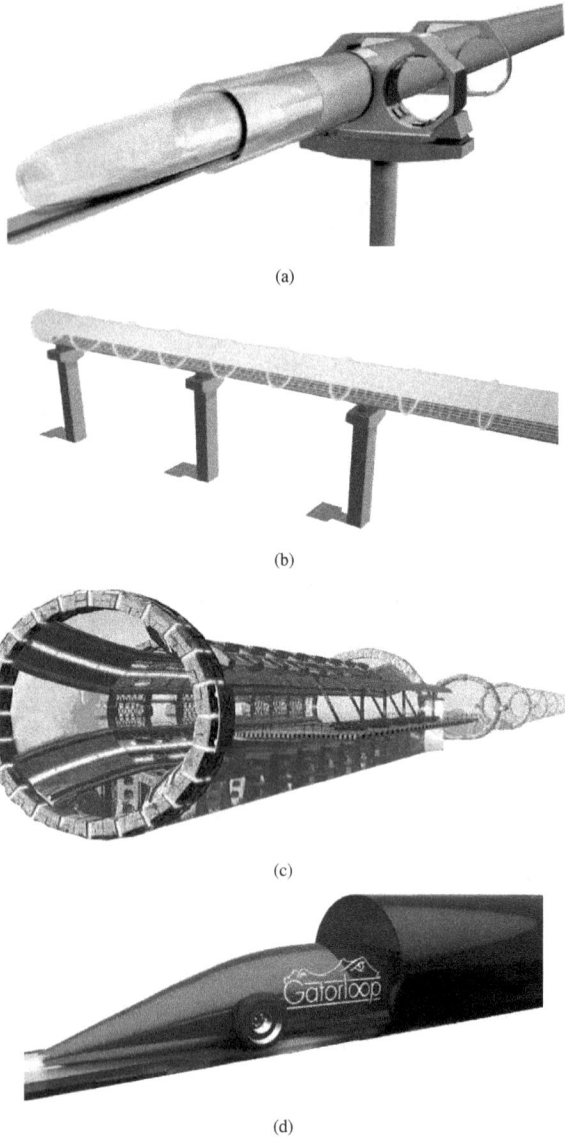

Figure 2.19   Simulation diagram of super pipeline. (a) Super line simulation diagram based on "steel shell pipeline." (b) Super line simulation diagram based on "transparent pipeline." (c) Super line simulation diagram based on "magnetic suspension pipeline." (d) Super line simulation diagram based on "vacuum pipeline."

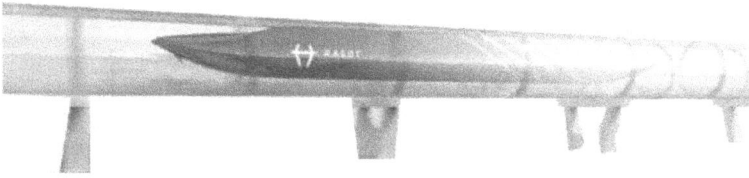

Figure 2.20   Linear shape of super pipeline.

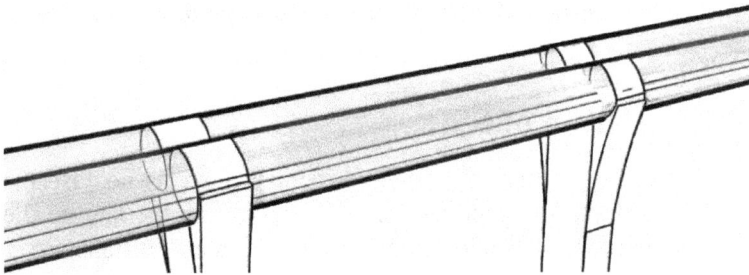

Figure 2.21   Elevated type of super pipeline.

gravity at most, that is, when the SSR train runs at a speed of 1,000 km/h, the turning radius must reach at least 65 km. Therefore, to ensure the comfort of passengers and the operational safety of the train, the SSR can only go in a straight line. The linear shape of the super pipeline is shown in Figure 2.20.

Requirements II: The "elevated pipeline" structure of the super line: To make the SSR go straight as far as possible and avoid disturbing residents, the ET3 Company proposed the "elevated pipeline" model (Figure 2.21). Like the familiar light rail, it supports the pipeline on a concrete pillar several meters high, which also greatly reduces the floor area and building area required on the ground. Thus, it can minimize the amount of land required by the line and the construction cost of the line.

Requirements III: The "mortgage seal" cabin transportation of the super line: Since the sealed cabin is guided directly to the pipe surface through air bearings and suspension, the SSR runs very smoothly and avoids the need for expensive rails. The sealed chamber of a super pipeline is shown in Figure 2.22.

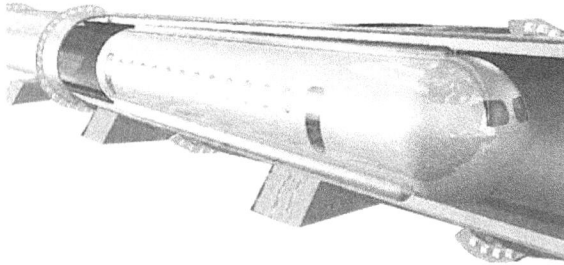

Figure 2.22 Sealed chamber of super pipeline.

2.3 Flying Train of SSR

"Speed is the eternal pursuit of human beings, there is no fastest only faster." Man's thirst for speed has never stopped from the land to the air. On land, humans are constantly developing new means of transport, from horse-drawn carriages at 20 km/h to steam trains at 70 km/h to wheel-rail high-speed trains at 350 km/h. "Speed" made the world focus on high-speed trains. But people want to go faster, so maglev trains are running at 500 km/h and flying trains at 1,000 km/h. Technology makes all imagination come true. The flying train can reach 1,000 km/h, and even exceed the speed of sound at 1,200 km/h. It is a new generation of transportation after ships, cars, trains, and airplanes, and will also be the fastest ground transportation in the world. The running speed of each vehicle is shown in Figure 2.23.

2.3.1 *Flying train*

On the one hand, the "flying train" is different from the airplane: It does not fly in the clouds and has no wings, and it runs fast on the ground. On the other hand, the "flying train" is different from wheel-rail vehicles: It does not travel along the track like a normal wheel-rail vehicle, but is suspended on the track.

Basic concept: The levitation train in the low-vacuum pipeline (Maglev & Low-vacuum mode), known as the "flying train," flies along the track of SSR. The flying train which speed is higher than 1,000 km/h is high-speed type, and those lower than 1,000 km/h are ordinary type.

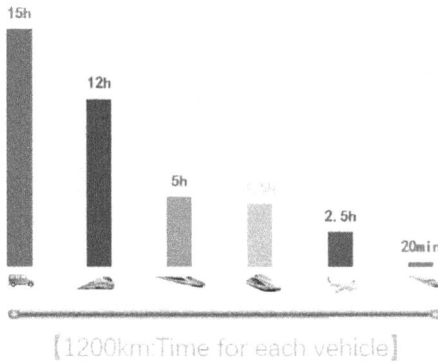

Figure 2.23   Running speed diagram of each vehicle.

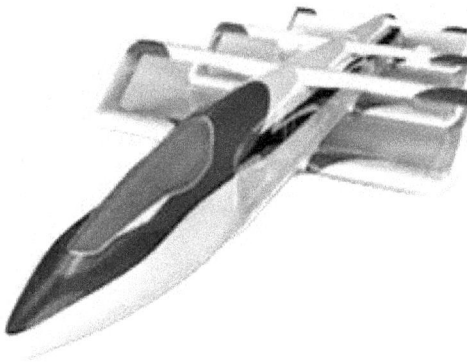

Figure 2.24   Japan's flying train.

For example, Tohoku University in Japan has developed a train with "wings" and hybrid aircraft features, as shown in Figure 2.24. Russian scientists have lifted an aircraft-style train off the ground to speed it up and speed over the ground, as shown in Figure 2.25.

Basic principle: The maglev train has no wheels and uses the electromagnetic force generated by superconducting maglev technology. The train floats on the track because it runs in a low-vacuum pipe with little wind resistance. Under electromagnetic thrust, the speed of the flying train is very great. For example, the design speed of the flying train on the Chinese test line has reached 1,000 km/h, which is faster than ordinary

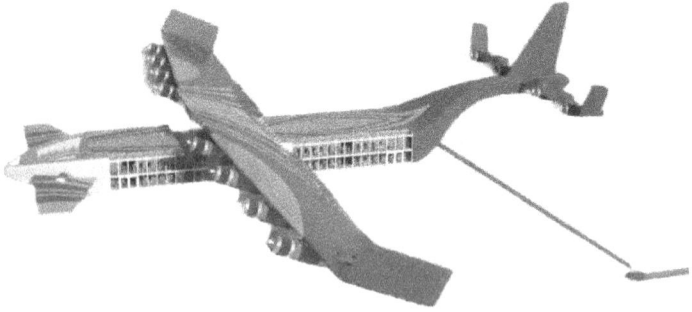

Figure 2.25   Russian flying train.

Figure 2.26   China's flying train.

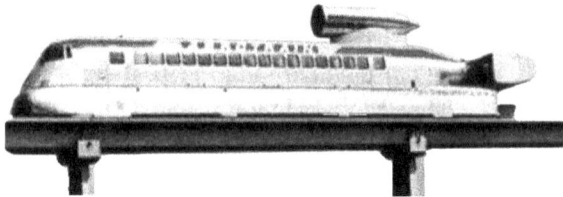

Figure 2.27   Flying train.

civil airliners and close to the speed at which a bullet is loaded, so it is called "high-speed flying train," as shown in Figure 2.26.

In a flying train, as shown in Figure 2.27, because the locomotive is not in contact with the ground, there is no friction to reduce its forward power, giving it a higher speed than conventional train locomotives, while reducing energy consumption. This revolutionary prototype of the train faces the same problems as all aircraft, where the wings tilt and wobble, making it difficult to balance the bottom with the ground. Therefore, although the flying train uses a built-in system to counteract the unwanted wing movement and stabilize the axis, it is not enough to run smoothly. Maglev trains use powerful electromagnets to hang the bottom of the body above the tracks, allowing the train to run above the tracks.

2.3.2 *Attributes characteristic of flying train*

The flying train has powerful ventilation, which lifts the train slightly, then starts the engine, and it's ready to go! Why is this the case? This is equivalent to an airbag, which produces the air supply function, the special device, and the ram-air device. The lower part of the train is fitted with a device that allows the train to lift slightly, up to 5–10 cm. This reduces friction, allowing the train to travel at a maximum speed of 600 km/h above the ground while using very little energy. High-speed flying trains have many advantages, such as not being affected by weather conditions, not consuming fossil energy, and being seamlessly connected to urban subways. As for the future "ticket price" of flying trains, the cost is changing. The larger the industry scale and the more mature the technology, the lower the cost will be. That's because flying trains cost only 25% of highways cost and 50% of regular high-speed train cost. In particular, the flying train has higher safety, lower energy consumption, less noise, and less pollution values than the general high-speed train.

Technical performance: Low vacuum + Supersonic + Magnetic levitation = Not affected by the weather. The flying train is a transportation system that uses a low-vacuum environment and supersonic shape to reduce air resistance, and magnetic levitation to reduce friction resistance to realize supersonic operation. The flying train not only shortens the space–time distance between cities but also has many advantages, such as not being affected by weather conditions, no consumption of fossil energy, and seamless connection with urban subways.

Operating speed: No resistance + Suspension + Solar energy = Fast operation. On the one hand, the flying train uses the low-vacuum environment and supersonic shape to reduce the air resistance; on the other hand, the flying train suspends the whole vehicle in the air and reduces the friction resistance through magnetic levitation to achieve supersonic operation. The whole vehicle is 20 mm away from the ground, using superconducting magnetic levitation technology and a vacuum pipeline to reach a maximum operating speed of 4,000 km/h.

2.3.3 Research and development vision

Countries all over the world have been watching the vast market of vacuum tube technology for a long time. At present, only the United States, Switzerland, China, Russia, Japan, Germany, France, and Canada, have made progress in the field of vacuum pipeline magnetic levitation technology. Given in the following are details of each countries progress:

(1) **Flying train in China:** A high-speed railway is an important means of transportation for people. Every year, hundreds of millions of passengers, from south to north, from east to west, embark on journeys to pursue their dreams with a vision of a better tomorrow. The rapid high-speed rail not only shortens the time and space distance between cities but also writes a new chapter on the land of China to consolidate the achievements of poverty alleviation, accelerate rural revitalization, promote economic and social development, and serve national unity and progress. The wheel of history rolls forward, and China's high-speed rail continue to march forward. With the dream of high-speed rail, we will move forward to the future together. However, the vast territory and the distance between the major urban centers is about 1,000 km. To create a national "one-hour economic circle," it is necessary to have vehicles of 1,000 km/h speed and high-speed cars. In this way, the dream of "flying close to the ground" will come true.

On August 30, 2017, China Aerospace Science and Industry Corporation Limited (CASICL) announced in Wuhan that it has launched a research and development project for a "high-speed flying train" with a speed of 1,000 km/h that flies near the ground, and will develop the SSR with a maximum operating speed of 2,000 km/h and a speed of 4,000 km/h. In September 2018, China's flying train project started carrying out key technology research, as shown in Figure 2.28. The flying train project aims to develop high-speed flying trains, reduce air resistance significantly through near-vacuum pipelines, and provide strong acceleration and high-speed cruise capabilities by using electromagnetic propulsion technology. China Aerospace Science and Industry Corporation (CASIC) announced recently that it will begin to develop a high-speed near-ground train with a maximum speed of 4000 km/h. If it is successfully developed, it will take less than 20 minutes to travel 1213 km/h from Beijing to Shanghai.

(2) **Flying train in Japan:** Japan has developed an aero-train with a speed of 400–500 km/h, which is the Japanese flying train, also

(a)

(b)

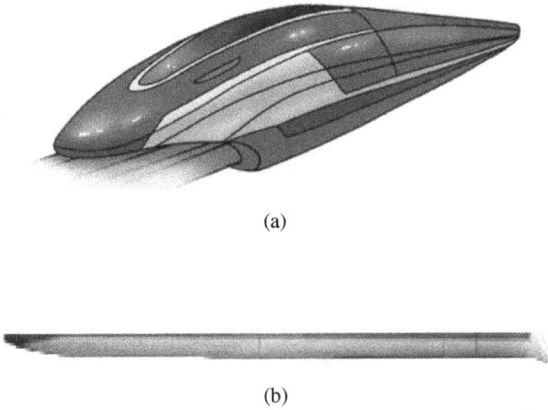

Figure 2.28   Flying train developed by China. (a) Maglev train. (b) Flying train.

known as a high-speed aero-train. Japan's flying train uses the aero-dynamic principle to realize the suspension of the new train, as shown in Figure 2.28. The flying train can run at a speed of 400–500 km/h, and its energy consumption is only about 1/6 that of ordinary maglev trains, so it can run entirely on clean energy. The flying train was developed by Tohoku University in 2001 and uses traditional wings to greatly improve the train's stability and lift-to-drag ratio. According to the Japanese plan, Japan will open a pneumatic hovertrain line with a design speed of 400 km/h by 2025. The second line will travel from Tokyo to Osaka at a speed of 500 km/h in less than an hour, compared to the current two hours.

The conductor of the Japanese flying train is more than one-meter long. On the one hand, the front of the flying train is like a bullet, and the body has annular wings and airflow thrusters. On the other hand, the flying train is somewhat similar to a plane, with wings. Japan's flying trains are powered entirely by natural energy, are cheap to use, and are very fast. Based on the speed of 500 km/h, the energy consumption of Japan's flying trains is 1/3 that of regular high-speed trains and 1/6 that of ordinary maglev trains. Japan plans to travel from Narita airport to Haneda airport via an underground tunnel with a designed speed of 400 km/h, which will connect the two airports in about 10 min, compared to more than an hour on the ground. The flying train developed by Japan is shown in Figure 2.29.

(a)

(b)

Figure 2.29   Flying train developed by Japan. (a) Flying train. (b) Track taxiing system.

(3) **Flying train in America:** The American flying train is also called the American Maglev aircraft, namely, MAGPLANE, as shown in Figure 2.30. The concept of Magplane was proposed by experts at the Massachusetts Institute of Technology. Magplane adopts permanent magnet electric suspension, linear synchronous motor drive, and electromagnetic turnout. Among them, the Magplane's suspension electromagnet and drive electromagnet are permanent magnets, with clearance up to 5~15 cm. On the one hand, the American flying train uses a 20-mm-thick curved aluminum plate track in suspension and guidance, which has the advantage of high-speed turning. On the other hand, Magplane's training wheels are fixed under the train and the overall technology is relatively simple.

At present, the American flying train has two applications: The first is a high-speed superconducting scheme with a speed of about 500 km/h. The second is a quasi-high-speed permanent magnet scheme with a speed of about 250 km/h. Among them, the quasi-high-speed

(a)

(b)

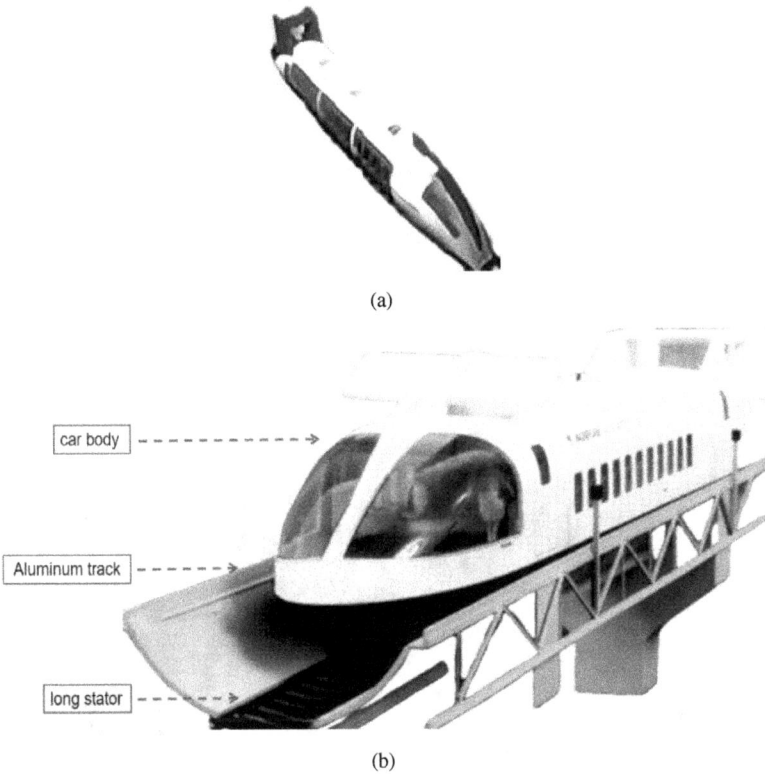

Figure 2.30 Flying train developed by American. (a) Flying train. (b) Track taxiing system.

scheme uses lower speed and smaller suspension clearance, and the flying train cost will be cheaper. Therefore, the Magplane's recent research and development have centered on this permanent magnet scheme.

(4) **Flying trains in Russia:** The Russian flying train is also called a surface plane or flying high-speed rail. Designed and developed by the Russian Concept company Dahir Insaat, the Russian Flying train is a futuristic, driverless high-speed train, as shown in Figure 2.31. The Russian-designed train runs at 600 km/h and can carry up to 4,000 passengers as it speeds across the surface. Russia's flying train is unmanned, changing the traditional understanding of conventional high-speed rail. Although the Russian flying train also has a certain

(a)

(b)

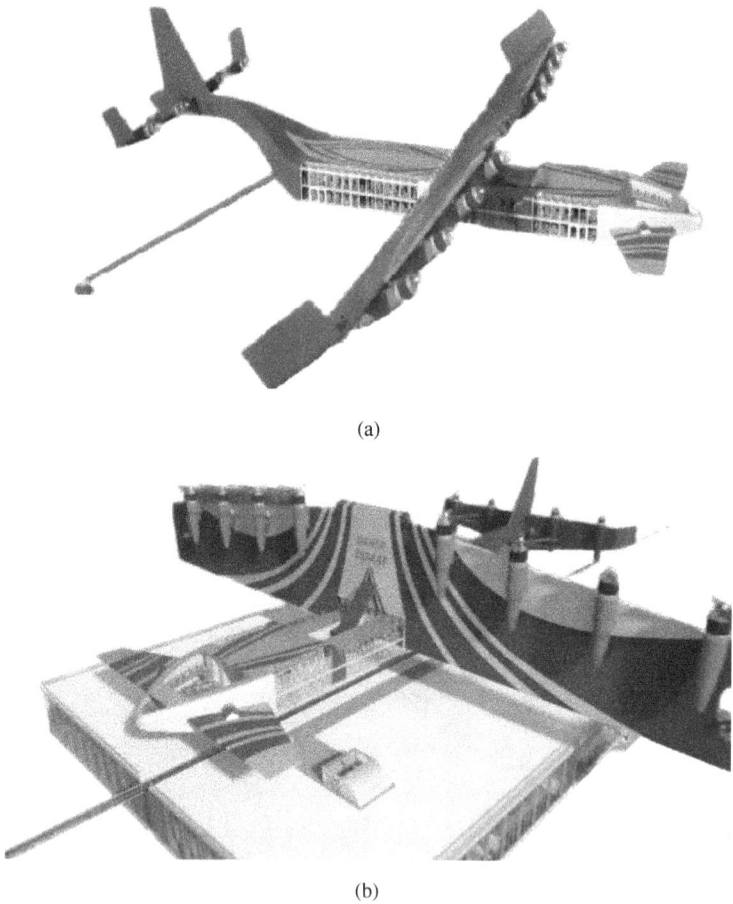

Figure 2.31   Flying train developed by Russia. (a) Flying train. (b) Track taxiing system.

track, mainly through a rope to fix the direction, to give people endless reverie, seems to be full of mystery! The Russian flying train is powered by a long tail attached to a track on the ground. The wing can change the angle and can realize vertical takeoff and landing. The flying train runs on purely electric power and can reach a cruising speed of 600 km/h. The carriages are designed with multiple layers, so they can seat more people than airplanes, with a capacity of about 1,000 to 4,000 people. The interior also features a restaurant and bar, as well as the requisite bathrooms. The seating arrangement in the interior also removes the barrier between people, making it more like a high-end restaurant built by a high-speed railway service rather than a high-speed train.

Figure 2.32   Big passenger train developed by Russia.

In the passenger compartment of the Russian flying train, which is only connected to the track by a power pole, the floating state of the flying train is derived from four boiler wheels and two small boiler turbines at the tail, which are also used to reduce resistance. At the same time, the flying train is completely off the ground, reaching a state of flight, and it does not need to occupy any space on the ground in the process of running. The biggest problem facing traffic is congestion, but this is not an issue for Russian flying trains. At the same time, the rail is made of wear-resistant and heat-resistant tungsten steel, which also has advantages. The Russian flying train is still in the conceptual design stage, and there are still many problems to be solved to realize it. For example, Russian flying trains can only fly on viaducts at a height of 5–6 m, and the flying height is only 5–10 m higher than the viaduct; the whole flying train is like a kite, connected to the rail by only a wire. If there is a strong wind, the "kite line" will break, and the people on board will be in danger. Although Russia's flying train also has tracks, it mainly uses electric poles to fix the direction of travel. This idea is full of imagination. A big passenger train developed by Russia is shown in Figure 2.32.

In short, today, it seems that the flying train will usher in an epoch-making traffic change, which will become the biggest breakthrough in railway and aviation in the world. In the future, there will be more subversive things in our lives. It is undeniable that human life will become more and more convenient and efficient. Life never favors those who are old-fashioned and satisfied with the status quo but leaves more opportunities to those who are brave and good at reform and innovation.

2.4 Summary

The transportation system has become the largest consumer industry of non-renewable resources and the main culprit for environmental pollution. Moreover, automobiles are the largest cause of casualties in the world, with more than 500,000 people killed in automobile accidents all over the world every year. Therefore, mankind must find a way out and explore new means of transportation with new theoretical knowledge, science, and technology. The SSR system (i.e., vacuum pipeline high-speed maglev transportation) has unique excellent performance, which is expected to fundamentally solve a series of problems faced by the transportation system.

Chapter 3

Basic Principle of Super-Speed Rail (SSR)

Elon Musk believes that the SSR system is relatively safe, and the system itself is not affected by complex weather. This is because, on the one hand, the SSR system can supplement energy by itself. After installing solar panels in the SSR system, the energy obtained will exceed the energy consumed by the whole system. On the other hand, the SSR system is also equipped with energy storage facilities, which can still drive for a week without using battery panels. The SSR system architecture is shown in Figure 3.1.

3.1 System Principle of SSR

Due to the speed and privacy offered by the pipeline, enterprises used to use the vacuum pipeline network to send newspapers between large buildings. The principle of the SSR system is based on the operation principle of the pipeline transportation system. Therefore, to reach the destination at high speed, the SSR system must be safe, fast, and comfortable. On the one hand, the super line should be long enough and must be maintained in a low-voltage environment, although it cannot be a real vacuum. On the other hand, to prevent the passenger compartment from touching the pipeline, the super-high-speed train will float slightly on the pipeline and run quickly. An operation simulation diagram of SSR system is shown in Figure 3.2.

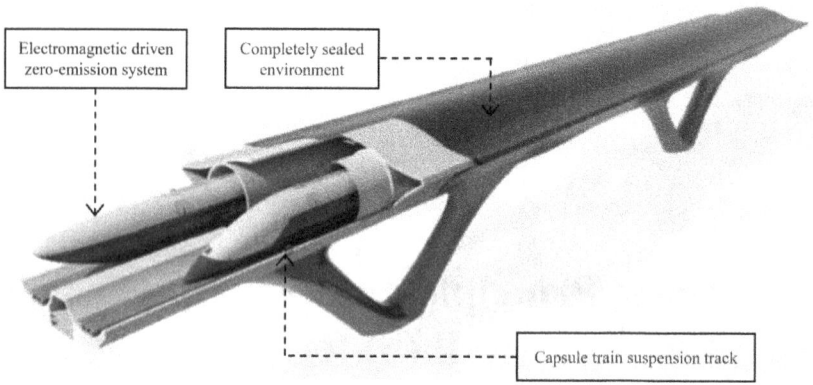

Figure 3.1 SSR system architecture.

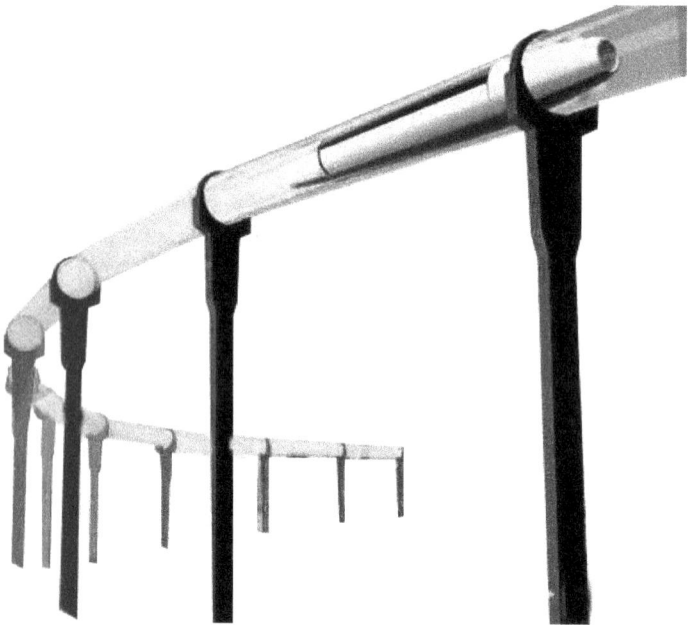

Figure 3.2 Operation simulation diagram of SSR system.

The SSR system is a kind of maglev train running in a vacuum pipeline. Since there is no need to solve the problem of excessive air resistance, the speed of the SSR system will be greater, and is expected to reach more than 1,200 km/h which is very fast.

Figure 3.3   Super line based on "solar power generation."

Figure 3.4   Super train pipeline model designed by Hyperloop.

(1) **The SSR system adopts clean energy:** The SSR system will adopt the self-powered design, and generate enough electric energy to maintain operation by laying solar panels on the upper part of the pipeline. At the same time, although evacuated tube transport can reach ultra-high speed, passengers will not feel the high-intensity acceleration. It will be safer, cheaper, and quieter than trains and planes. Evacuated tube transport consumes less energy than any other traditional means of transportation. The transportation volume of evacuated tube transport per kilowatt-hour is twice that of trains, which also reduces the dependence of transportation tools on fossil fuels and greenhouse gas emissions. A super line based on "solar power generation" is shown in Figure 3.3.

(2) **The SSR system adopts a vacuum pipeline:** The interior of the pipe is evacuated to reduce air resistance. In addition, the compression fan device at the front end of the super train can suck in air and discharge it from the bottom of the train to form an air cushion several millimeters thick, to suspend the train and reduce friction consumption. Figure 3.4 shows the super train pipeline model designed by Hyperloop.

3.2 Basic Principle of Super Train

The design of the super train in the SSR system does not take the traditional train carriage as a reference. The capsule transport cabin is the means of transport, and the "capsule" is placed in the steel pipe and then launched to the destination like a shell. The capsule transport cabin runs continuously in an environment with almost no friction. According to Elon Musk at the 2013 Science and Technology Conference, the speed of the super train will be about two or three times that of the bullet train or about twice that of the aircraft. According to the person in charge of the Hyperloop SSR construction project, the maximum speed of the upcoming SSR for testing is 700 km/h, while the speed of the fully completed SSR will reach 1,220 km/h. Magnetic levitation acceleration devices configured everywhere in the pipeline push the transport cabin forward. Figures 3.5 and 3.6 show the super train track model designed by the Hyperloop.

Just as the aircraft climbs to a high altitude and crosses the air with low density (the relationship between altitude and pressure is shown in Table 3.1), the SSR system encapsulates the train in a pressure relief pipe to avoid using a hard vacuum, because a vacuum environment is expensive and difficult to maintain for a long time compared with the low-pressure solution. Despite the low pressure, aerodynamic challenges still need to be addressed. These measures include controlling the formation of shock waves when the speed of the sealed cabin is close to the speed of sound and the air resistance increases sharply. Near cities where many

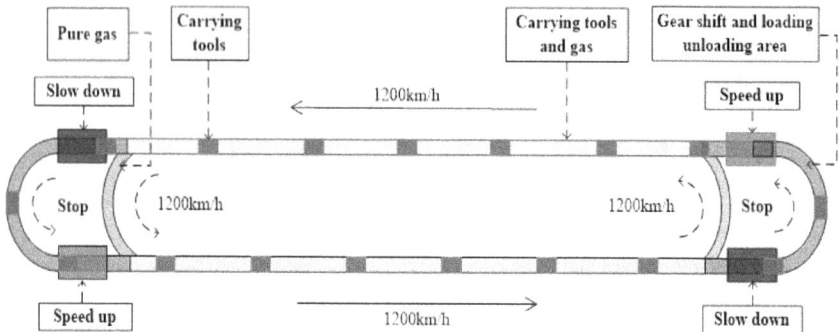

Figure 3.5   Operation diagram of SSR system.

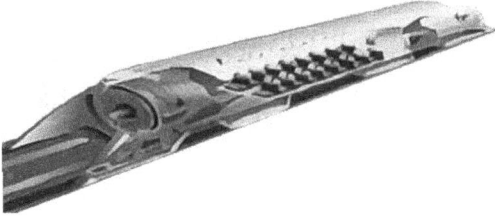

Figure 3.6 Super train track model.

Table 3.1 Relationship between altitude and air pressure.

h (m)	P (kPa)	h (m)	P (kPa)	h (m)	P (kPa)
0	101.3	4,000	61.6	8,000	35.6
1,000	89.9	5,000	54.0	9,000	30.7
2,000	79.5	6,000	47.2	10,000	26.4
3,000	70.1	7,000	41.0	11,000	22.6

turns are required, the capsule needs to travel at a slower speed, which reduces the perceived acceleration of passengers and also reduces the power requirements of the capsule.

3.3 Basic Principle of Super Station

The super station is an indispensable part of the SSR system. The reasonable setting of the station plays a vital role in the fast and smooth operation of the whole system. Station construction should take into account three aspects: the station location, the operation mode of the station, and the construction requirements of the station. The setting of super stations is different from the existing HSR stations. The Wuhan HSR station is shown in Figure 3.7 which shows Wuhan Station Hierarchy Structure.

(1) **Location selection of the super station:** Based on the operation characteristics of the SSR system, the super station will not occupy too much building area. In addition, due to the low noise during the operation of the SSR, the super station can be located in the city

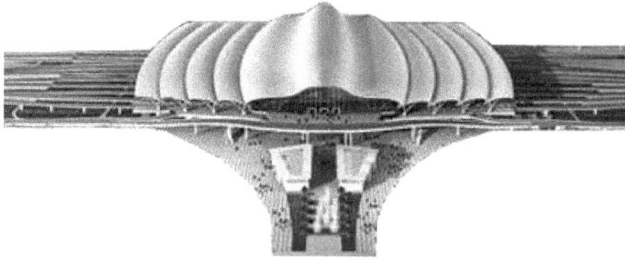

Figure 3.7   Wuhan HSR station.

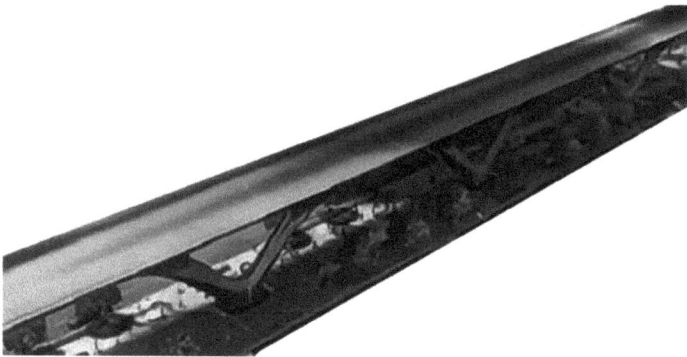

Figure 3.8   Location map of super station.

center, which will be more convenient for passengers (Figure 3.8). According to the "zero-distance" transfer requirements, the station planning and construction should be a conducted in comprehensive way that organically connects it with other transportation modes. Megacities should strengthen the convenient connection of SSR passenger transport hubs, airports, and urban rail transits.

(2) **Operation characteristics of the super station:** All ticketing and baggage tracking of SSR will be processed electronically, and passengers do not need to print boarding passes and baggage tags. Because the SSR makes the travel time very short, this mode of transportation is mainly used for commuting rather than vacations. Each person will be limited to 2 bags of luggage, with a total of no more than 50 kg. Luggage will be placed in separate compartments at

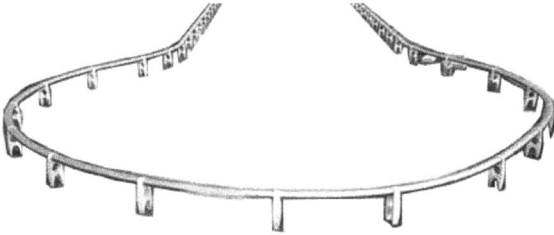

Figure 3.9   Operation diagram of super station.

Figure 3.10   Transfer diagram of super station.

the front and rear of the transport cabin in a manner similar to the high-altitude dustbin on the passenger plane. The luggage compartment can be taken out from the cabin so that the process of loading and retrieving luggage can be carried out separately from the cabin arrival and departure. In addition, the staff of the super station will be responsible for loading and unloading passenger luggage to maximize the operational efficiency of the train. (Figures 3.9 and 3.10 show the envisaged super transfer process.)

(3) **How a super station works:** The transfer area of the SSR terminal will be a large open area with two large airlocks, indicating the entry point and exit point of the transport cabin. The arriving transport module will enter the airlock, where the pressure will be balanced with the space station and then released to the transit area. The doors on both

Figure 3.11 Working diagram of SSR station.

sides of the transport compartment will open to allow passengers to get off. The super station staff will quickly unload the baggage compartment or separate it from the transport compartment so that baggage claim will not interfere with the turnover of the transport compartment. When the passengers and luggage leave the transport cabin, the transport cabin will rotate and align on the turntable to reenter the operation pipeline as shown in Figure 3.11.

3.4 Basic Principle of Super Line

The suspension cabin in the SSR system can be suspended in the pipe (Figure 3.10). Its lifting force is generated by the air in front of the vehicle inhaled and compressed by the fan and compressor on the left side, and then expelled from a group of 28 air-bearing skids under the vehicle body. These skids are isomorphic with the geometry of the pipe wall. Each of them is 1.5 m long and 0.9 m wide. They support the weight of the suspended cabin by floating on a pressurized air pad 0.5–1.3 mm above the ground. The air pressure under the skid only needs to reach 9.4 kPa (about 9% of the sea level air pressure) to support the suspended cabin. Working diagrams of the SSR pipeline are shown in Figures 3.12 and 3.13.

In the SSR system, Elon Musk's Hyperloop company proposed air cushion suspension, a representative support method for vacuum pipeline trains, and other research institutions use magnetic levitation as a suspension support scheme for vacuum pipeline trains. Simulation diagrams of American SSR are shown in Figures 3.14 and 3.15.

Figure 3.12 Working diagram of China's SSR pipeline.

(a)

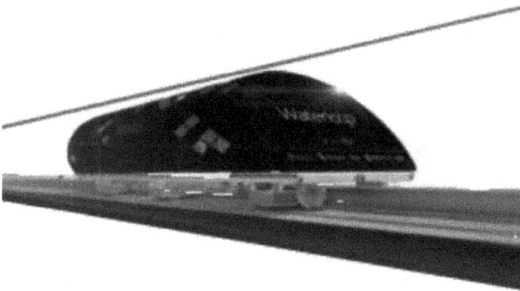

(b)

(c)

Figure 3.13 Working principle diagram of super pipeline. (a) Internal section of super pipeline. (b) Exterior drawing of super pipeline. (c) Internal operation of super pipeline.

Figure 3.14   Simulation diagram of American SSR.

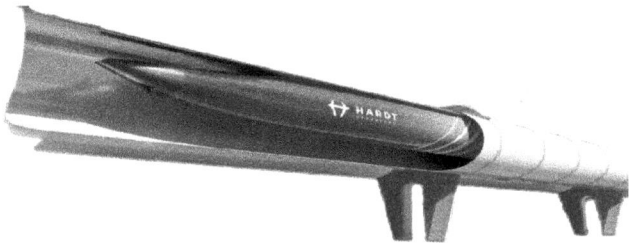

(a)

(b)

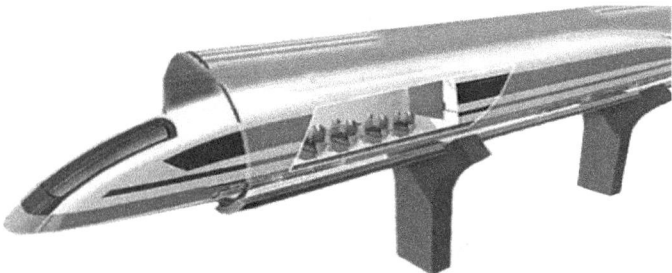

(c)

Figure 3.15   Simulation diagram of American SSR. (a) Internal operation of SSR. (b) Internal operation of SSR. (c) Internal operation of SSR. (d) Internal operation of SSR. (e) Internal operation of SSR.

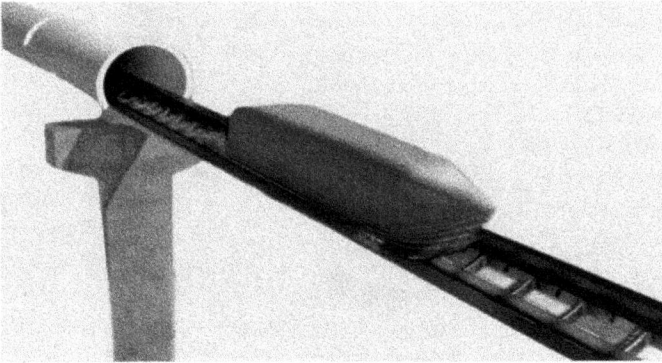

(d)

(e)

Figure 3.15 (*Continued*)

3.5 Summary

The SSR system is a revolutionary mode of transportation. According to the design, the "capsules" cabin will be installed in a fixed vacuum tube system on the ground that acts like a railway track, allowing it to move quickly under the power of the sun. Although the idea of the SSR system was put forward by Elon Musk, CEO of Tesla Motors and SpaceX, Hyperloop Technologies is one of the enterprises that is trying to make it a commercial operation. Therefore, the SSR system from an idea to the realization and then to the final practical application, are inseparable from the design and practice of scientists.

Chapter 4

System Architecture of Super-Speed Rail (SSR)

The SSR system is mainly composed of the super trains, super stations, and super lines. The super train in the SSR system is a passenger cabin, which is shaped like a capsule, so it is also called "capsule train." The super line in the SSR system is different from the line of the HSR. The SSR line is a closed pipeline, and the interior of the pipeline is in a vacuum, to eliminate air friction. The super station in the SSR system is mainly for passengers to embark and disembark, and its layout is much simpler than the HSR station and the airport.

4.1 System Composition of SSR

The SSR system is a complex system. Macroscopically, it is mainly composed of super trains, super lines, and super stations. Microscopically, it is mainly composed of the super train microarchitecture system, the super line architecture system, and the super station architecture system. The organizational structure of the SSR system is shown in Figure 4.1.

4.2 Organization Structure of Super Train

The super train in the SSR system is a capsule car running in a low-pressure pipeline. Elon Musk said that it is the fifth mode of transportation after aircrafts, trains, ships, and automobiles. It takes into account the characteristics of speed, economy, and safety, and is also an open-source

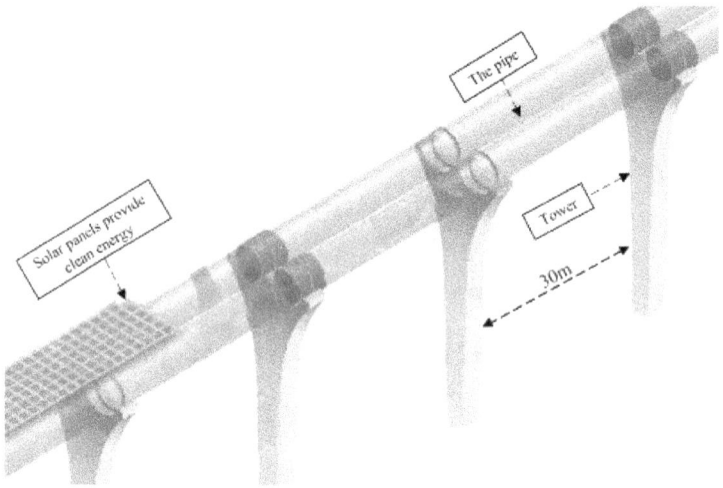

Figure 4.1   Organizational structure of SSR system.

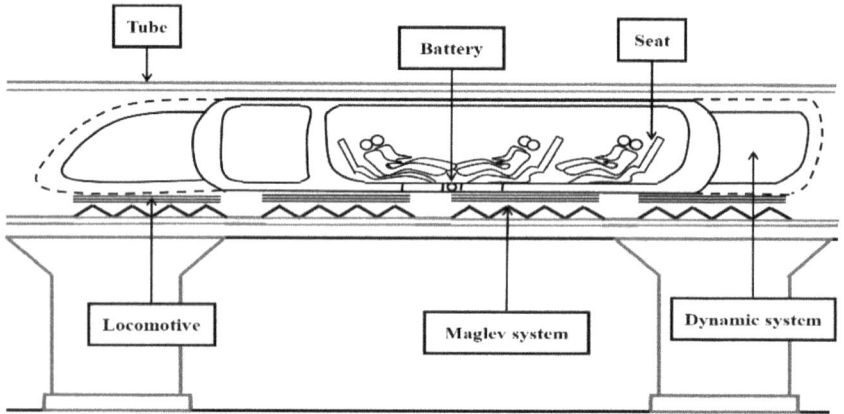

Figure 4.2   Organization chart of super train.

solution. For example, Elon Musk's SSR was originally designed for intercity traffic between Los Angeles and San Francisco, but it is also applicable to intercity traffic with a spacing of less than 1,500 km. The organizational chart of a super train is shown in Figure 4.2.

The super train shown in Figure 4.2 has a capacity of 28 passengers, a departure frequency of 3 min, and an average speed of 1,200 km/h.

There are two passengers in a row, and the seats are equipped with a personal entertainment system. The luggage is concentrated in the front or rear of the sealed cabin. When the whole system is accelerating continuously, the acceleration borne by passengers is 0.5 g (half of the gravity).

4.2.1 *External structure of super train*

From the current research situation, the SSR system essentially uses magnets to provide thrust and compressed air to provide lift. In this design, the SSR system will not have friction resistance between the "wheel and track." At the same time, to reach the target speed per hour, low pressure should be maintained in the running track to reduce the resistance between the train and the air. The super train in the SSR system will use a small cabin and can start at any time without following a fixed timetable like an aircraft. The schematic diagrams of several super trains under development are shown in Figure 4.3(a)–(c). Figure 4.4 shows the car shape of the super train, Figure 4.5 shows the head of the super train, and Figures 4.6 and Figures 4.7 show the shell skeleton of the super train.

The super train is a vehicle of the SSR transportation system, which is divided into two types: passenger super train and passenger-cargo super train. Super trains in the SSR system start every 2 min, every 30 s in peak hours, and the distance between adjacent trains is 37 km. The maximum speed that the super train can reach is 1,220 km/h, and it will slow down when approaching the city. The present speed comparison of various vehicles is shown in Figure 4.8.

The main structure of the super train in the SSR system includes a compressor, water tank, and air-bearing sliding plate as detailed in the following:

(1) **Compressor of the super train:** A compressor is installed in the front of the super train in the SSR system. On the one hand, it can suck in the air in front of the train to eject it from the rear of the train to prevent the flow from causing a blockage between the train and the pipe wall when passing through the narrow pipe. On the other hand, air bearings can be formed to support the train.

(2) **Water tank of super train:** The super train in the SSR system is also equipped with a water tank to reduce the air temperature, and the tail is equipped with a battery to supply power for the equipment required by the train, such as the compressor and refrigerant.

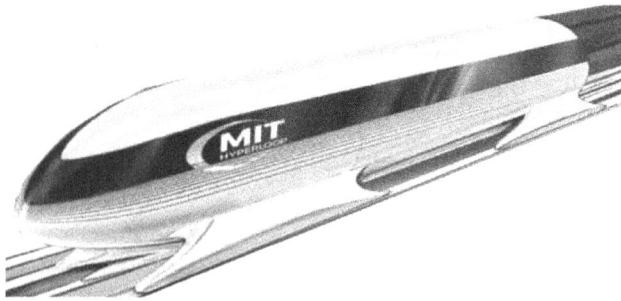

(a)

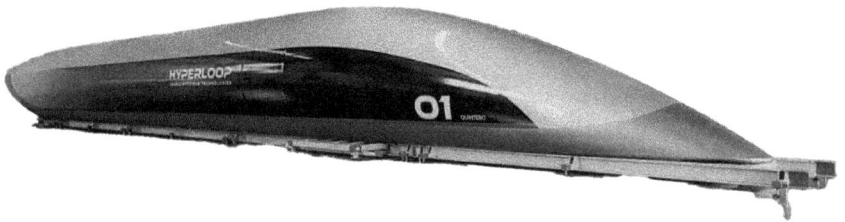

(b)

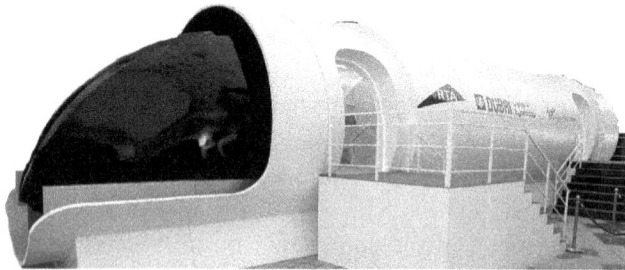

(c)

Figure 4.3   Model diagrams of super train in different companies.

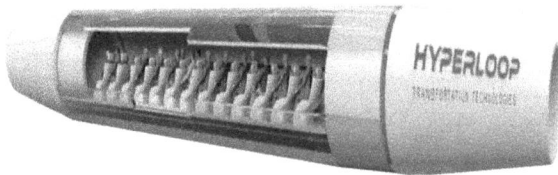

Figure 4.4   Interior diagram of super train.

Figure 4.5   Outline of Hyperloop train carriage.

Figure 4.6   Design of Hyperloop train head.

Figure 4.7   Hyperloop train shell skeleton.

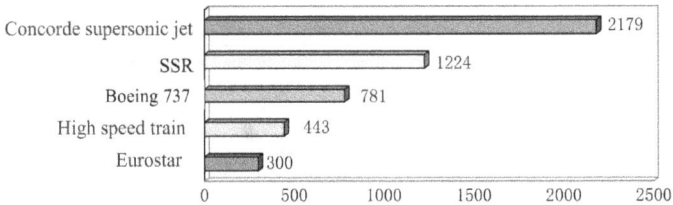

Figure 4.8   Running speed of various vehicles (km/h).

(3) **Air-bearing sliding plate of the super train:** There is an air-bearing sliding plate supporting the pod at the bottom of the super train in the SSR system. When the interval between the sliding plate and the pipe wall is reduced, the airflow will form a strong pressure between them to make them return to the original height. To alleviate the discomfort caused by this lifting and bumping, the sliding plate is also equipped with a suspension device.

4.2.2 *Internal structure of super train*

According to Elon Musk's expectation, the transportation system in the SSR includes a low-pressure steel pipe and aluminum capsule body. The interior of the cabin is protected by gas, and the maximum operating speed will exceed 1,220 km/h. Because there are no windows in the cabin and the space is narrow, the designers use a virtual reality screen on the cabin top to broadcast some pictures similar to blue sky and white clouds, creating a sense of openness, to eliminate the claustrophobia of passengers. In addition, the super train also has business class and even private office areas. The SSR system is designed to provide passengers a pleasant and impressive trip. The internal structure of the super train is shown in Figure 4.9.

The overall internal weight of the transport cabin in the SSR system is expected to be close to 2,500 kg, including seats, restraint systems, door

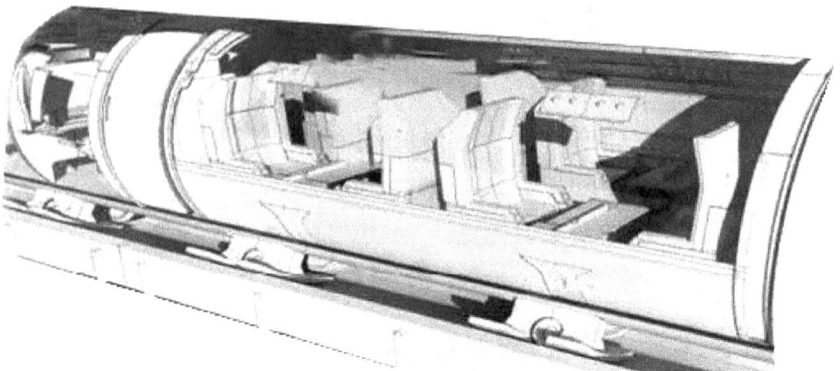

Figure 4.9   Interior diagram of Argo design's SSR.

panels, luggage compartments, and entertainment displays as detailed in the following:

(1) **Seats for super trains:** The interior of the super train in the SSR system is specially designed to take into account the safety and comfort of passengers, and the seats are ergonomically designed. Although the vacuum transport pipeline can reach an incredible speed, passengers can only feel a small explosive acceleration, and stay comfortable and safe during the high-speed acceleration. The seats of the super train are shown in Figure 4.10.

(2) **Windows of super trains:** Augmented reality windows will be used in the transport cabin of the SSR system, which can not only show the real world outside the windows but also add digital information with entertainment attributes. The windows of the super train are shown in Figure 4.11.

(3) **The luggage compartment of the super train:** In order to facilitate the rapid loading and unloading of passengers' luggage by the staff at the super station in the SSR system, the luggage compartment is set at the front and rear ends of the train. The passenger and luggage compartments of the super train are shown in Figure 4.12.

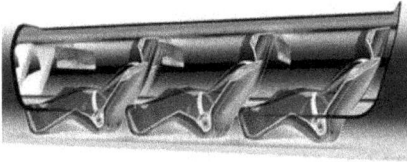

Figure 4.10   Seats of super train.

Figure 4.11   Window of super train.

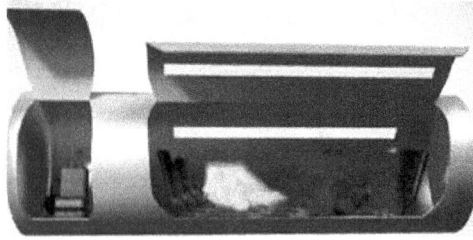

Figure 4.12 Passenger compartment and luggage compartment of super train.

(4) **The transport cabin of the super train:** As the super train in the SSR system operates in the vacuum pipe and the transport cabin is pressurized, each seat is equipped with an oxygen mask during train operation to ensure the safety of passengers.

(5) **Ventilation system of the super train:** To make passengers comfortable and safe and ensure the air quality in the vehicle, the super train in the SSR system is equipped with an air exchange system. There are three schemes to eliminate air pollution in the super train compartment: dilution method, adsorption method, and separation method.

4.2.3 *Equipment composition of super train*

In addition to installing equipment and facilities for passenger needs, super trains in the SSR system also include airborne compressors, on-board power supply, suspension equipment, and propulsion equipment that play important roles in train operation. This book will introduce some equipment parameters of passenger super trains and passenger-freight super trains in the following:

(1) **Airborne compressor of SSR system:** The airborne compressor is a piece of important equipment to ensure the operation of the super train in the SSR system. It has two purposes: On the one hand, the airborne compressor compresses the air bypassing the train, so that no air flow blocks the space between the capsule and the pipe wall, allowing the super train to pass through relatively narrow tubes. On the other hand, it also provides air to the air bearing to support the weight of the capsule throughout the journey.

(2) **On-board power supply of SSR system:** In addition to the compressor motor and coolant, the passenger compartment power system in

the SSR system also includes a battery with an approximate weight of 2,500 kg to provide sufficient power.

(3) **Suspension system of SSR system:** Because the super train in the SSR system needs an ultra-high-speed operation, suspending the super train in the pipe is a major technical challenge. The SSR system utilizes air-bearing suspension, which provides stability and extremely low resistance at a feasible cost by using the air in the pipeline as shown in Figure 4.13.

(4) **Propulsion system of SSR system:** To propel the vehicle at the target speed, the designers are developing an advanced linear motor system to accelerate the transport cabin to more than 1,220 km/h with up to 1 g of acceleration. The moving motor element (rotor) is placed on the

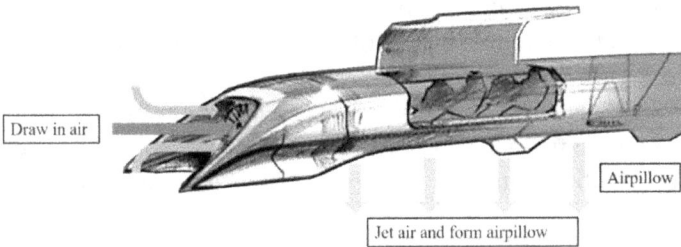

Figure 4.13   Suspension system of super train.

Figure 4.14   Hyperloop train power drive.

vehicle to reduce weight and meet power requirements, while the fixed motor element (stater) providing power for the vehicle will be placed on the pipeline. Figure 4.14 shows the hyperloop train power drive.

4.3 Organization Structure of Super Station

The boarding process and station layout of the SSR system are much simpler than boarding at the airport, so the design concept of the super station is extremely simple and practical. With short travel times and frequent departures, safety and security are paramount as each super station is expected to have a continuous flow of passengers. The security inspection method of the super station in the SSR system will be similar to that of the airport. This process will be optimized procedurally, to reduce the waiting time and maintain a more sustainable passenger flow. The super station is shown in Figure 4.15.

The SSR system is a high-speed passenger transport system, which connects major cities and urban areas with high speed and high frequency. The super stations in the SSR system operate at two levels and structures:

(1) **The first floor of the super station: Transition hall:** The transition hall in the SSR system on the first floor is used as auxiliary transportation for boarding and alighting passengers to take them to other public transport systems.

(2) **The second floor of the super station: Entrance and exit hall:** The entrance and exit halls in the SSR system on the second floor are the main areas for passengers to embark and disembark the super train. Passengers enter and leave the super train through the aisle.

Figure 4.15 Conceptual design of super station.

(3) **The internal auxiliary structure of the super station: Shaft and elevator:** The shaft and elevator in the SSR system are used to connect two levels. To enable passengers to enter and exit the super station hall smoothly, the cavern space outside the tunnel needs to be set on the second floor.

(4) **External auxiliary structure of the super station: Gate:** The SSR system is equipped with the longitudinal automatic gate within the station. To avoid vacuum loss, the super train must pass through the gate chamber before stopping at the station. When the train stops at the correct position in the station, the semi-vacuum tunnel gate will be closed automatically, so that passengers will be under normal air pressure in the hall when getting on and off, and normal air pressure will be maintained in the super train as well.

4.3.1 *Location selection of super stations*

Based on the operation characteristics of the SSR, the station in the SSR system cannot occupy too much building area. In addition, due to the low noise characteristics of the SSR operation, the super station can be located in the city center, which is more convenient for passengers. According to the "zero distance" transfer requirements, the planning and construction of super stations should be such that passenger can easily and organically connect with other modes of transportation.

4.3.2 *Operation mode of super station*

All ticketing and baggage tracking of the SSR system will be processed electronically. Passengers do not need to print boarding passes and baggage labels, so the boarding process is very simple. When the super train arrives at the super station, the doors on both sides of the transport cabin will be opened to let passengers get off. The station staff will quickly unload the luggage compartment or separate it from the transport cabin. After the passengers and luggage completely leave the transport cabin, the super train will reenter the operation pipeline. The passenger compartment and luggage compartment of the super train are shown in Figure 4.12, and the bus driving to the super train station is shown in Figure 4.16.

Figure 4.16   Super train station.

4.3.3 *Construction requirements of super station*

The super station in the SSR system is mainly the area where passengers buy tickets, accept security inspection, check in, or claim luggage. In addition to supporting service facilities, the superstation shall also meet the following requirements:

(1) **Accuracy of super station:** The super station in the SSR system enables safe and punctual operation of super trains in the SSR system.

(2) **Security of super station:** The super station in the SSR system can carry passengers and cargo safely and quickly, and is also comfortable for passengers.

(3) **Supply of super station:** The super station in the SSR system can maintain and supply the super trains.

(4) **Convenience of super station:** The super station in the SSR system enables passengers and cargo to reach nearby destinations smoothly. An outline drawing and an internal diagram of a super station are Figures 4.17 and Figures 4.18, respectively.

4.4 Organizational Structure of Super Line

The super line in the SSR system is composed of pipelines. In addition to various problems inside the pipeline, route planning is also a challenging task. In the research, some scholars found that passengers can bear the lateral force of one-tenth of gravity at most when the train turns, that is,

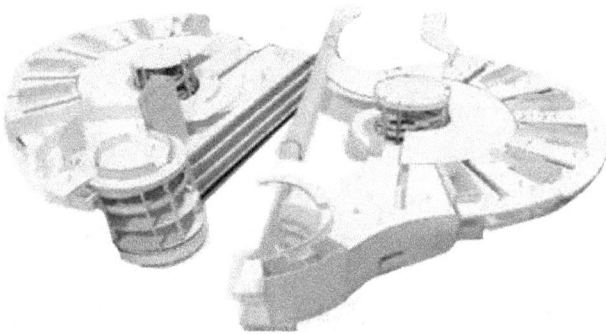

(a)

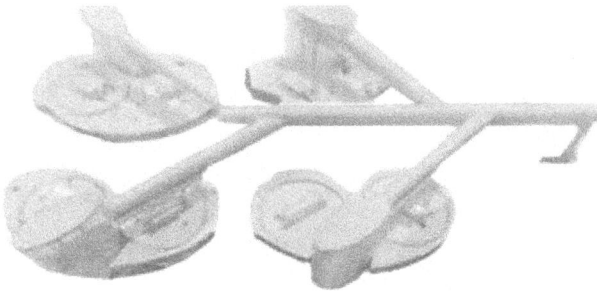

(b)

(c)

Figure 4.17 Outline drawing of super station in different companies. (a) Outline drawing of super station based on "arc." (b) Outline drawing of super station based on "five lines." (c) Outline drawing of super station based on "bat."

(a)

(b)

Figure 4.18 Internal diagram of super station. (a) Aerial perspective. (b) ground perspective.

when the super-high-speed train runs at a speed of 1,000 km/h, the turning radius should reach at least 65 km. Therefore, to ensure the comfort of passengers and the safety of train operation, the SSR system can only go in a straight line. The vacuum pipeline of the SSR system is shown in Figure 4.19 and the operation of the super train in the vacuum pipeline is shown in Figure 4.20.

4.4.1 *Geometric structure of super line*

The whole pipeline in the SSR system is composed of a steel structure, and the two pipelines will be welded together in a side-by-side manner so

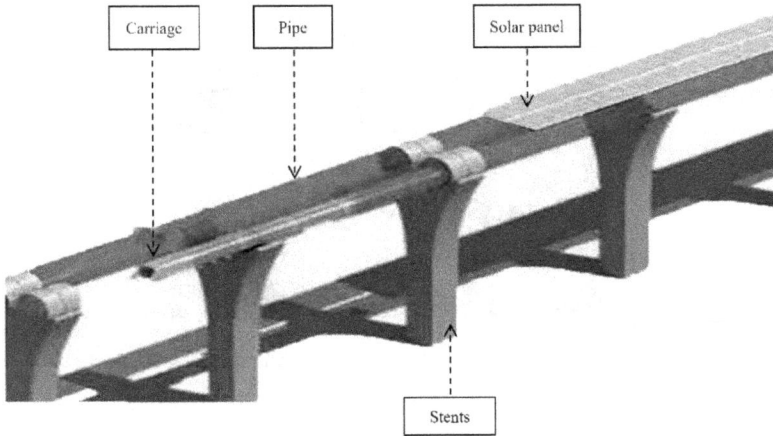

Figure 4.19 Operation of super train in vacuum pipeline.

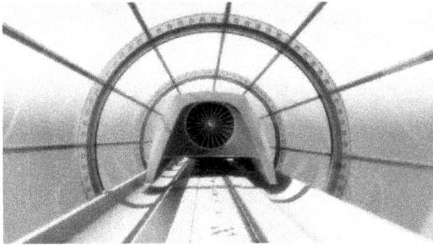

Figure 4.20 Operation of super train in vacuum pipeline.

that the transport cabin can operate in both directions at the same time (Figure 4.21). There will be a tower every 30 m in the SSR system to support the operating pipeline. At the top of the pipe, solar panels will be fixed to power the entire SSR system. The expected pressure in the SSR system pipeline will be maintained at 100 Pa, about 1/1,000 of the earth's atmospheric pressure. This low pressure minimizes the resistance on the transport cabin and can extract air relatively easily from the pipeline.

The geometry of the vacuum pipe in the SSR system depends on the passenger version of the super train or the passenger-cargo version of the super train. If the air accelerates to supersonic speed through the gap, shock waves are formed, and these fluctuations limit how much air can discharge around the capsule. With the increase of resistance and air quality, the power demand of the capsule increases significantly. Therefore, it

Figure 4.21   Inside of vacuum pipe.

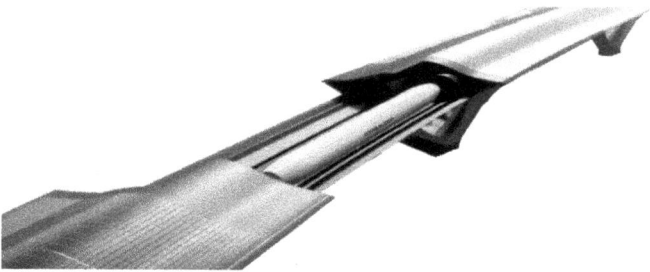

Figure 4.22   Operation of super train in vacuum pipeline.

is very important to carefully select the area ratio of the capsule to the pipeline to avoid the formation of shock waves around the capsule, which can ensure that the air around and through the capsule is sufficient at all operating speeds, and any air that cannot pass between the capsule and the pipe can be bypassed by the on-board compressor in each capsule. The operation of a super train in a vacuum pipeline is shown in Figure 4.22.

4.4.2 *Structural design of super line*

The vacuum pipeline in the SSR system is meant to further improve the speed of the super train by removing the dense atmosphere under the surface environment. Therefore, building a sealed pipe to evacuate the air in the pipe and maintain the required vacuum is the basis of the whole evacuated tube transport system. The problems of the vacuum materials, bearing structure, and sealing characteristics of the vacuum pipeline in the SSR system need to be solved urgently. The vacuum pipeline of the existing load seal separation scheme is shown in Figures 4.23 and 4.24. The structural design of the super line is described in the following:

(1) **Vacuum pipe of SSR system:** The vacuum pipeline in the SSR system mainly includes the underground pipeline and overhead pipeline.

    (i) *Underground pipeline of SSR system*: The underground pipeline in the SSR system buries the vacuum pipe underground, and is regarded as an extension of tunnel technology. However, the implementation cost of this method is high, the progress is slow, and it is not convenient to set up an emergency rescue system deep underground. The operation of the super train in the vacuum pipeline is shown in Figure 4.25.

    (ii) *Overhead pipeline of SSR system*: The vacuum pipe in the SSR system is erected in the air, just like the current HSR line. However, this method will expose the pipeline to the atmospheric

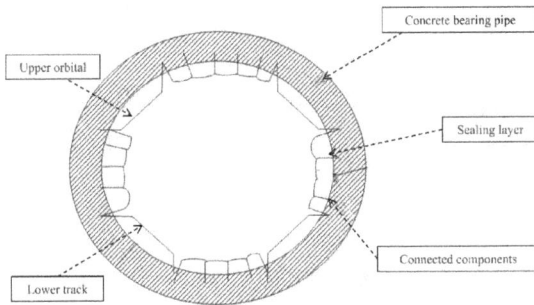

Figure 4.23   Schematic diagram of vacuum pipeline in load seal separation scheme.

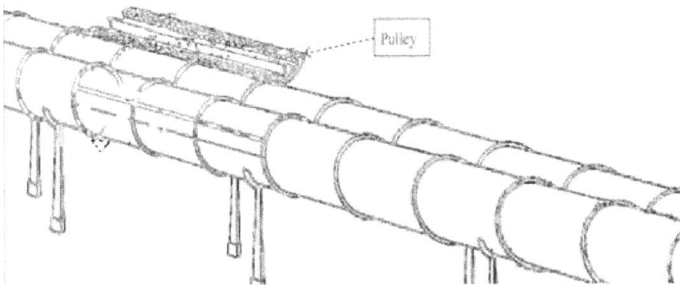

Figure 4.24   Pipeline system architecture.

Figure 4.25　Operation of super train in vacuum pipeline.

Figure 4.26　Operation of super train in vacuum pipeline.

environment, which is greatly affected by meteorological changes, causing it to interact with various complex situations on the ground. The curve of the ground limits the speed of the super train. The operation of the super train in the vacuum pipeline is shown in Figure 4.26.

(2) **Vacuum degree in the pipeline in the SSR system:** The vacuum is set to 1% in the SSR system. The setting of the vacuum degree in the SSR system should be related to the speed and power. The resistance of the object moving in the air is directly proportional to the second power of its moving speed, as shown in Equation (4.1), and the power consumption is directly proportional to the third power of its moving speed, as shown in Equation (4.2):

$$f = k\rho S v^2 \tag{4.1}$$

$$N = k\rho S v^3 \tag{4.2}$$

where ρ is air density,

v is relative air velocity along the moving square of the train,

S is vertical maximum windward cross-sectional area,

k is resistance coefficient, related to streamlining, generally determined by a test.

When the super train in the SSR system runs at the speed of 800 km/h, the power consumption to overcome the residual air resistance in the super train line will be maintained at the power consumption level of ordinary trains (under standard atmospheric pressure) running at 100 km/h. Due to the increase of speed, the operation time of super trains will be shortened, and the total energy consumption of super trains will be reduced. Considering the pipe effect, when other conditions are the same, it is necessary to consider selecting a higher vacuum degree. The air pressure of the vacuum pipe (Figure 4.27) in the SSR system should be maintained at 1/1,000 of the atmospheric pressure. From the perspective of vacuum technology, such a vacuum degree belongs to the low vacuum range.

(a)

(b)

Figure 4.27   Field installation of SSR system. (a) Field installation of vacuum pipeline. (b) Field installation of SSR train.

Table 4.1 Performance characteristics of common building materials.

No.	Material	Advantages	Disadvantages
1	Glass	Good air tightness	Low stress intensity
2	Metal	Good air tightness, high stress intensity	Cost is too high
3	Film, plastic	Good air tightness, cheap	Soft and unstable
4	Concrete	High stress strength and high cost performance	Air tightness is poor

(3) **Vacuum pipe material for SSR system:** The vacuum pipeline materials in the SSR system, that is, the materials constituting the vacuum environment boundary, need to have six basic characteristics: bearing, sealing, baking venting, chemical stability, thermal stability, and processing performance. The basic load to be borne by the vacuum pipeline in the SSR system is the pressure formed by the air pressure difference inside and outside the pipeline, and the pressure formed by the air pressure is perpendicular to the pipeline surface. In addition to the pressure formed by the air pressure difference, if the vacuum pipeline in the SSR system is buried underground, it also needs to bear geotechnical pressure and self-weight. The airtightness of the materials means that the vacuum pipe has the function of preventing gas penetration to maintain the stability of the internal vacuum environment. Table 4.1 shows the performance characteristics of common building materials.

(4) **Pipe section of SSR system:** The Japanese superconducting maglev train is 3.28 m high and 2.9 m wide when suspended. The German permanent magnetic levitation train is 4.06 m high and 3.7 m wide when suspended. The ratio of the cross-sectional area of the maglev train to the cross-sectional area of the pipeline is about 0.12, which is less than that of the wheel rail HSR. The reason is that the speed of the maglev train is higher, which will produce a domestic air-blocking effect in the tunnel. To increase the transport capacity and facilitate the fast exchange of passengers, the vehicle body running in the vacuum pipeline in the SSR system is considered to adopt the single-layer double-channel or double-layer double-channel design. According to rough estimation, the height or width of the super train will reach 6 m. In addition to the factors such as track bed height and suspension gap, the inner diameter of the vacuum pipeline should be

Figure 4.28   Simulation structure diagram of the superloop.

at least 9 m, which can be reached using the existing tunnel technology.

(5) **Towers and tunnels of SSR system:** The pipeline in the SSR system will be supported by columns, which restrict the pipeline in the vertical direction, but allow the longitudinal sliding of thermal expansion and inhibit the transverse sliding to reduce the risk caused by earthquakes. In addition, the nominal position of string connections in the SSR system will be adjustable vertically and laterally to ensure correct alignment in case of possible land subsidence, and these minimum-restraint column pipe joints will make the whole journey smoother. The simulation structure of the superloop is shown in Figure 4.28.

4.5 Summary

This chapter introduces the architecture of the SSR system in detail, including the super train, super line, and super station. First, the SSR system, the internal and external structure equipment, and vehicle type design of the super train are introduced. Second, in relation to the super station in the SSR system, the chapter focuses on the operation mode, location selection, and construction requirements of the station. Finally, with regard to the super line in the SSR system, the chapter mainly introduces the principle and structural design of the vacuum pipeline.

Chapter 5

Attribute Characteristics of Super-Speed Rail (SSR)

The SSR system is the "fifth mode of transportation" after ships, trains, automobiles, and aircraft, and is an ideal model of transportation with a speed of 1,000–2,000 km/h. Although it is still in the research and development stage, its envisaged performance is better than the HSR system and flying system. The operation of the super train in the closed vacuum pipeline is not affected by air resistance, friction, and weather, especially the natural environment (such as strong wind, rainstorm, debris flow, and low temperature). Therefore, in addition to the advantages of speed, the SSR system also has the characteristics of comfort, safety, stability and reliability, all-weather operation, economy, and environmental protection. The operation diagram of the SSR system is shown in Figure 5.1.

5.1 Basic Properties of SSR

The attributes of the SSR system are mainly reflected in its speed and transportation capacity. In terms of speed, the SSR system is a super train running in a vacuum pipe, and the speed per hour exceeds several times the flight speed of the aircraft. In terms of transportation capacity, according to Elon Musk's assumption, each super train can accommodate 28 people, but due to its small departure interval, the transportation capacity of the SSR system can fully meet people's travel needs.

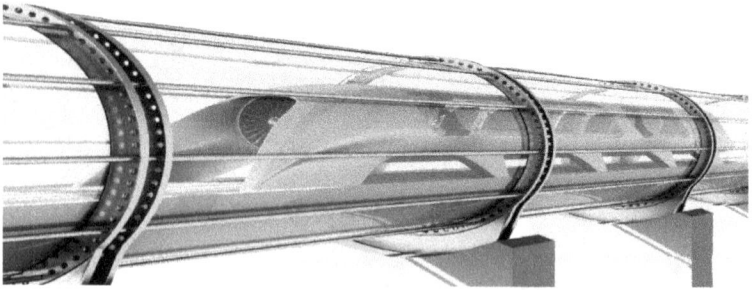

Figure 5.1   Operation diagram of SSR system.

5.1.1 *Speed of SSR*

The SSR system is the fastest transportation system. The running speed of the SSR is higher than 1,200 km/h, and it is also the fastest vehicle at present. Whether it is the king on the ground (HSR system) or the master in the air (aviation system), the operation speed of other modes cannot be compared to that of the SSR system:

(1) **The speed of HSR system:** High-speed vehicles will be affected by air resistance and contact friction during operation. The maximum speed of surface vehicles is about 500 km/h, while the maximum speed of a pipeline transportation system can reach more than 20,000 km/h in theory. The following are examples:

First, air resistance is at 300 km/h stage: When the speed of HSR exceeds 300 km/h, the main resistance comes from the air, and the air resistance exceeds 80%.

Second, air resistance is at 400 km/h stage: When the speed of HSR reaches 400 km/h, the resistance from the air exceeds 90%.

Third, air resistance is at 500 km/h stage: When the speed of HSR reaches 500 km/h, the resistance from air exceeds 99%.

When the vehicle runs on the ground, it faces atmospheric pressure. The HSR system is the king of ground speed. The operating speed limit is 400 km/h. Of course, it can run at 486.1 km/h, 574.8 km/h, and 605 km/h, but these are not economic speeds. The slow speed of the HSR is mainly affected by ground air resistance. Figure 5.2 shows the global HSR network as of 2019. Figure 5.3 shows the global HSR network in 2030.

(2) **The speed of the aviation system:** The main reason why the speed of HSR is lower than that of aircraft is that the air resistance

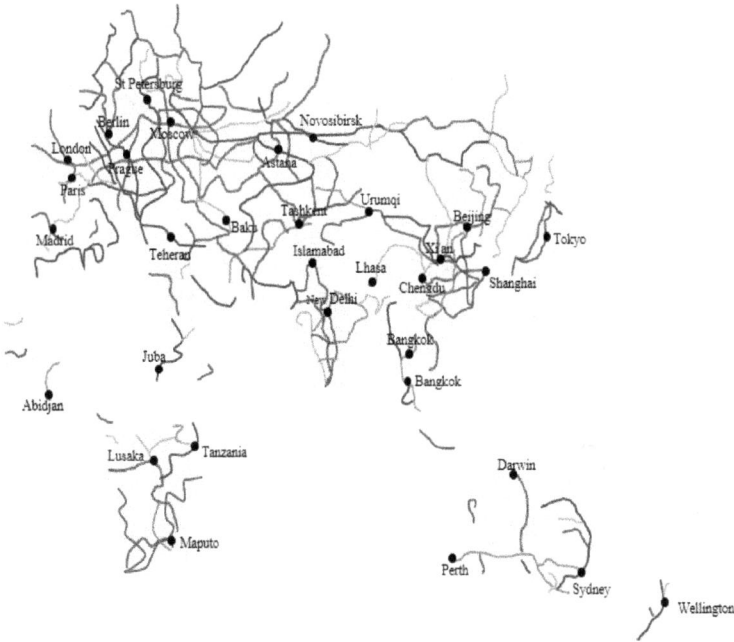

(a)

(b)

Figure 5.2 Global HSR network in 2019. (a) HSR network in the eastern hemisphere. (b) HSR network in the western hemisphere.

(a)

(b)

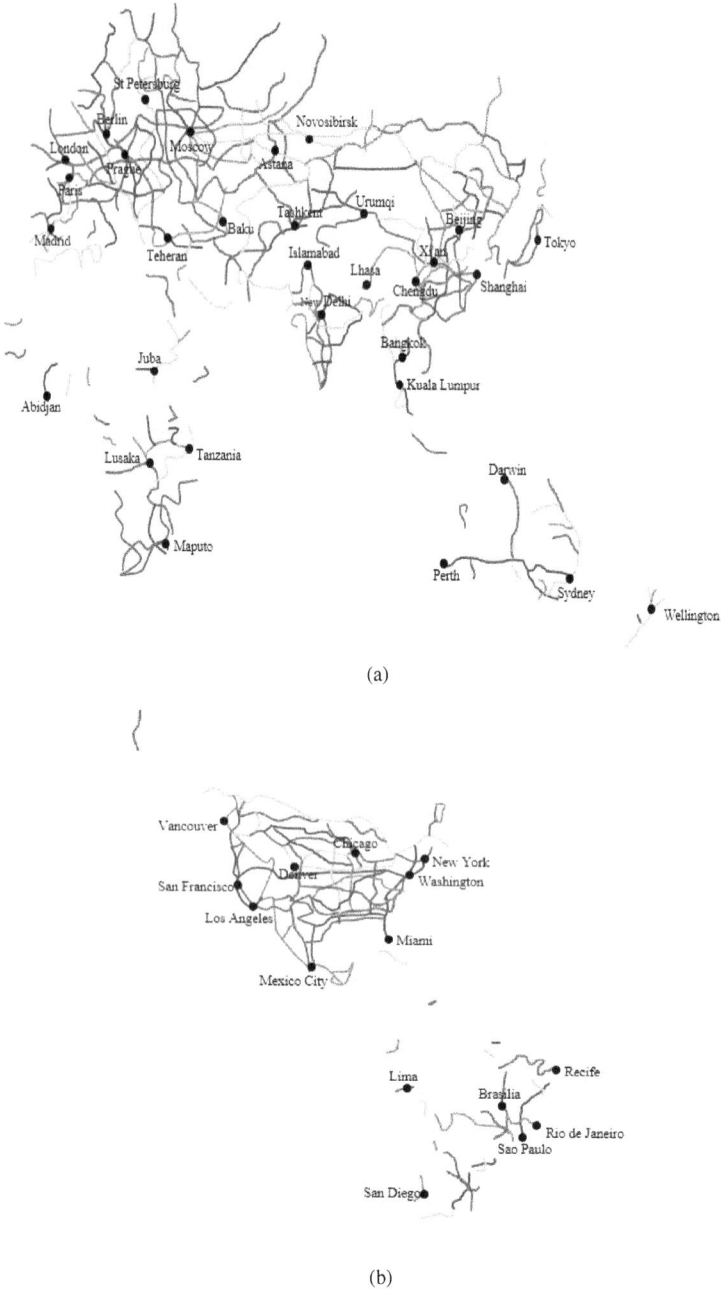

Figure 5.3 Global railway network in 2030. (a) HSR network in the eastern hemisphere. (b) HSR network in the western hemisphere.

Table 5.1 Optimum speed values at different atmospheric pressures.

Height (m)	Atmospheric pressure (Pa)	Flight speed (km/h)	SSR operation speed (km/h)
<1,000	1	400–500	<500
1,000–4,000	0.8–1	500–600	500–1,200
4,000–10,000	0.5–0.8	600–800	
10,000–12,000	0.2–0.5	800–1,000	1,200–2,000
12,000–15,000	0.05–0.2	1,000–2,000	
15,000–20,000	0–0.05	2,000–10,000	2,000–20,000
>20,000	0	>10,000	>20,000

encountered by HSR is far greater than that encountered by aircraft. The aircraft flies higher than the HSR; the higher the altitude, the thinner the air, and of course, the smaller the resistance, so the speed of the aircraft is higher than that of the HSR. The relationship between air density and height is as follows:

First, the low-altitude stage: 4,000–6,000 m: At the altitude of 4,000–6,000 m, there is only 0.5 atm, which is the altitude used by regional aircraft. Its economic speed is between 400 and 800 km/h.

Second, the hollow stage: 6,000–15,000 m: At an altitude of 10,000 m, there is only 0.2 atm. This is the altitude of trunk aircraft, and the economic speed is 800–1,000 km/h.

Third, the high-altitude stage: above 15,000 m: At an altitude of 15,000 m, there is only 0.05 atm. It is the world of ultra-high-speed aircraft, and the economic speed can reach 2,000 km/h.

According to the results of existing research, the optimal speed values under different atmospheric pressures are shown in Table 5.1.

5.1.2 *Transportation capacity of SSR*

Transportation capacity is an important attribute of various means of transportation. At present, the vehicle with the largest transportation capacity is the HSR, whose carrying capacity is 10 times that of air and 5 times that of expressways (the comparison between the carrying capacity of HSR and other vehicles is shown in Table 5.2). Statistics show that the maximum passenger volume of an expressway will not exceed 10 million person-times a year, while the passenger volume of an HSR system will

Table 5.2   Comparison of carrying capacity between HSR and other means of transport.

Type of shipping	HSR	Expressway	Transport aviation
Carrying capacity (based on HSR)	1	1/5	1/10
Transportation cost (based on HSR)	1	2.5	5

reach 150 million person-times a year. For example, in Japan's Tokaido Shinkansen, an average of 11 trains depart per hour during peak hours, and 283 trains pass through every day. Each train can carry 1,200–1,300 passengers, with an annual passenger transport of 120 million. The maximum hourly two-way transportation capacity of HSR, expressway, and aviation is shown in Table 5.2.

In the SSR system, a super train needs to be arranged with 28 seats, which seems to be a small carrying capacity. However, due to the extremely fast operation speed of the super train, the departure interval is generally about 2 min, and the fastest one can reach 30 s in the peak period. 40 super trains will be configured in the peak period, and six super trains will be available for passengers to get on and off at the terminal at the same time. In this way, the transportation capacity of the SSR system is not as large as that of the HSR, but since the SSR system is mainly used for long-distance transportation, such transportation capacity is still considerable and can fully meet people's travel needs.

5.2   System Characteristics of SSR

The SSR system is a vehicle designed based on the theory of "Evacuated Tube Transport," which adopts maglev technology. Its design concept and operation principle determine the operation safety, passenger comfort, system stability, vehicle reliability, and economy of the SSR system.

5.2.1   *Safety of SSR*

No matter what kind of transportation, safety is the first priority. Therefore, safety and reliability are the primary factors for passengers to travel. The SSR system uses huge and almost vacuum pipelines to connect multiple cities to form the SSR network, which is convenient for everyone to travel quickly. Safety has been considered in the design of the SSR system from the beginning. The system is not easily affected by natural disasters, human factors, derailment accidents, etc., as mentioned in the following:

(1) **The SSR system is not vulnerable to natural disasters:** Natural disasters have a great impact on various means of transportation. Many major traffic accidents are caused by natural disasters. The types of natural disasters mainly include earthquakes, gales, lightning, rainstorm, and debris flow. Ships, trains, cars, and aircraft will be affected by natural disasters. In comparison, the SSR system has obvious advantages in this regard. The SSR system is a train running in a vacuum pipeline (Figure 5.4), and the vacuum pipeline is a completely closed system, isolated from the external environment and not affected by bad weather such as strong wind, earthquakes, lightning, and rainfall. The comparison between the safety of the SSR system and other transportation modes is shown in Table 5.3.

(2) **SSR system is not susceptible to human factors:** Among the many factors threatening traffic safety operations, human factors are undoubtedly the most noteworthy. Most traffic accidents, especially road related, are caused by human factors. According to statistics, motor vehicle traffic accidents in China account for 92.7% of total accidents every year. The different human factors causing road traffic accidents are shown in Table 5.4.

The impact of human factors on train accidents cannot be ignored. According to statistics, 30.5% of train accidents are caused by direct human error and 68.6% are caused by equipment failure due to human error. In addition, among the factors affecting aviation safety, the primary role is also human factors. The statistics show that more than

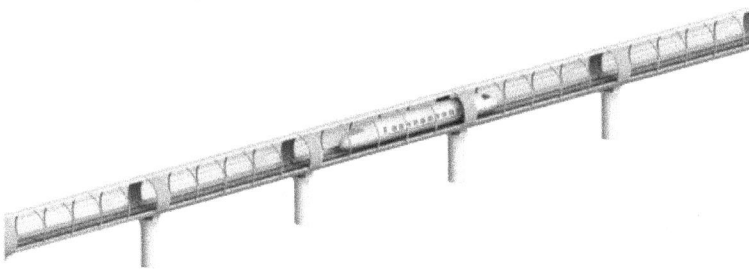

Figure 5.4 Operation of SSR in pipeline.

Table 5.3   Impact of natural disasters on various transportation modes.

Mode of transportation	Earthquake	Temperature	Thunder	Rainstorm	Debris flow
SSR	Minimum	Minimum	Minimum	Minimum	Minimum
Aircraft	Little	Minimum	Maximum	Little	Little
Train	Maximum	Great	Great	Maximum	Maximum
Automobile	Great	Maximum	Little	Great	Great

Table 5.4   Human factors affecting road traffic accidents.

No.	Factors affecting drivers' ability to deal with accidents	No.	Factors inducing drivers to take risks
1	Lack of experience	7	Overestimate your ability
2	Alcohol and drug abuse	8	Disobey traffic rules
3	Drowsiness and fatigue	9	Inappropriate driving behavior
4	Sudden disease	10	Habitual overtaking
5	Severe psychological stress	11	Alcoholism
6	Temporary distraction	12	Taking psychiatric drugs

Table 5.5   Statistics of flight accidents and error types determined in accident characteristics.

No.	Error type	Percentage (%)
1	Skill	29
2	Knowledge, regulations	26
3	Statute	21
4	Knowledge	10
5	Knowledge and skills	7
6	Regulations and skills	7

80% of flight accidents are caused by human error as shown in Table 5.5 and Figure 5.5.

From Table 5.5 and Figure 5.5, it is found that most traffic accidents are related to human error, and the SSR system operates in a completely closed vacuum pipeline, which has a fully automatic

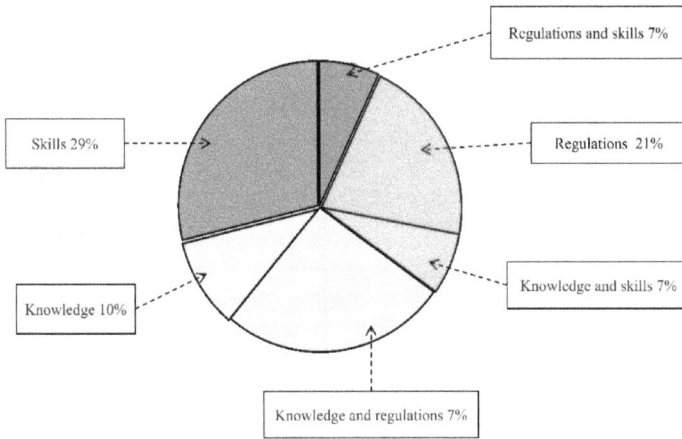

Regulations and skills 7%

Skills 29%

Regulations 21%

Knowledge 10%

Knowledge and skills 7%

Knowledge and regulations 7%

Figure 5.5 Statistics and proportion of error types determined in flight accidents and accident characteristics.

control system. There are no mistakes of drivers, pedestrians, staff, and managers in the operation of the SSR system. This advanced intelligent control system has unmatched safety compared to other modes of transportation and tools. In addition, safety cabins are also set at intervals along the vacuum pipeline of the SSR system. When the super train stops due to failure or the pressure in the sealed cabin is lost, passengers can escape to the safety cabin to avoid danger.

(3) **There is no derailment in the SSR system:** Both ordinary trains and HSR run on the track, so derailment can occur for various reasons, resulting in serious casualties (Figure 5.6). Some train derailment accidents in the world in recent years are shown in Table 5.6.

The SSR system operates in a vacuum pipeline, adopts magnetic levitation technology, and does not need a track. Therefore, there is no derailment in the operation of the SSR system. In short, whether objectively or subjectively, if compared with vehicles such as cars, aircraft, ships, and wheel rail HSR, the SSR system is very safe. Once put into operation, the SSR system will be the safest vehicle. A Hyperloop SSR train is shown in Figure 5.7.

5.2.2 *Comfort of SSR*

Comfort is another key reason why people choose the SSR system for travel. The internal design of the SSR system takes into account the

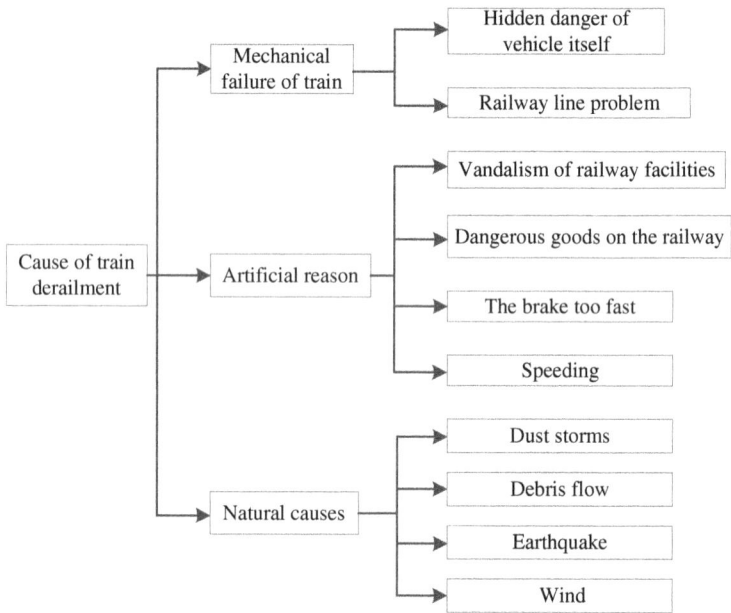

Figure 5.6   Causes of train derailment.

Table 5.6   Train derailment accident.

Time	Country	Number of deaths	Number of injured
June 3, 1998	Germany	101	105
April 25, 2005	Japan	107	549
July 23, 2011	China	42	192
November 24, 2015	Spain	79	180
July 14, 2016	Italy	27	50

comfort of passengers: On the one hand, although the speed of the SSR system can reach more than 1,000 km/h, passengers can fully adapt to this speed. On the other hand, the seats inside the SSR system are well in line with the human body to maintain the comfort of passengers during travel. The SSR system will use augmented reality windows to solve the problem of viewing the scenery without glass, and the beautiful landscape will be displayed in the cabin. Each passenger can use his entertainment system

Figure 5.7 SSR train of hyperloop.

to make the journey enjoyable. In addition, passengers can hardly feel the impact of noise during the operation of the SSR system. The comfort of the passengers can be ensured with the following:

(1) **Passengers can well adapt to the vacuum environment of the SSR system:** The super train in the SSR system runs in a vacuum pipe. Will the human body not adapt to this vacuum environment? Rob Lloyd, President of Hyperloop One, said the following: "When people ride the SSR, they sit in the suspension-sealed cabin. Although the track is in a vacuum state, the suspension-sealed cabin can be pressurized. Therefore, we don't have to worry about the impact of the vacuum environment on the human body." Others believe that the speed of the SSR system is faster than that of the aircraft, and the human body will not be able to bear it, when in fact, according to scientific analysis, the human body can handle such speeds.

Comparative analysis of the SSR system and automobiles based on human adaptability indicates that the human body's maximum tolerance for acceleration is usually around 50 m/s^2, while a car's 100 km acceleration is about 10 s. If the SSR can accelerate to 1,000 km/h within 1–2 min, it is not a problem for the human body to bear the running speed of the SSR system. Moreover, for passengers, although the speed is very fast, they will not feel the high-intensity acceleration in a vacuum environment.

Comparative analysis of the SSR system and aircraft based on human adaptability indicates that the acceleration of the SSR system can reach the acceleration of the aircraft. The acceleration of the aircraft is generally 0.5–0.6 g, which is equivalent to about 5 m acceleration per second. The acceleration of people running 100 m is much greater than this. This acceleration is no problem compared with ordinary people. When the super train is operated at a certain speed, it will move in a uniform manner. At that time, passengers will not feel any speed, just like astronauts flying in space. Therefore, if the acceleration of the super train in the SSR system is too fast, people can still stand it.

(2) **The seats of the SSR system are more comfortable:** There are supposed to be about three seats in a row for the SSR system. The front and rear seats are very spacious, and they are all leather seats. The seat angle can also be in line with the body. It does not look inferior to China's HSR. According to the previous idea of Elon Musk, a super train may take up to 28 people. The seat assumption diagram of the SSR system is shown in Figure 5.8. Figure 5.9 shows the seat assumption of the SSR business class.

(3) **The SSR system makes little noise:** The noise generated by transport vehicles will cause annoyance and harm to people living and working near the transport route. Moreover, traffic noise limits the potential use of space along the line, which may lead to the opportunity cost of land use. The influence of noise depends on the noise level of the noise source, the number of people exposed to the noise, and the duration of noise exposure, which means that in the considered system, this performance mainly depends on factors such as transportation

Figure 5.8   Seat assumption of SSR.

Figure 5.9   Seat assumption of SSR business class.

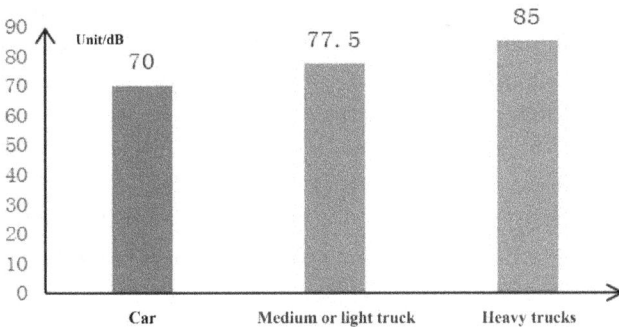

Figure 5.10   Comparison of average noise generated by different vehicles (unit: dB).

route, speed, and the number of passing vehicles. The noise emitted by different modes is described in the following:

(i) *Car noise*: The noise generated by driving on the road comes from the vibration noise of the motor vehicle engine shell, air inlet sound, exhaust sound, horn sound, braking sound, and the noise formed between the tire and road surface. When the motor vehicle is running at low speed, the vibration and noise of the engine shell are loud. When running at high speed, the tire noise rises. The measurement results show that when the speed is 50–100 km/h, 15 m away from the center of the traffic trunk line, the minibus is 65–75 db loud, the medium or light truck is 70–85 dB loud, and the heavy truck is 80–90 dB loud. The speed doubled, and the traffic noise increased by 7–9 dB on average. The average noise generated by different vehicles is shown in Figure 5.10.

(ii) *Railway noise*: Railway traffic noise includes signal noise, locomotive noise, and wheel-rail noise. The signal noise varies due to
the steam pressure used by the whistle or the compressed air pressure used by the whistle. At 10 m from the side of the locomotive,
the whistle sound can reach 120–140 dB, and the sound level of
the whistle is about 30–40 dB higher than locomotive noise.
Locomotive noise mainly includes electric locomotive noise, diesel locomotive noise, and steam locomotive noise. The noise level
of the electric locomotive is the lowest, the noise level in the cab
is about 82–87 dB and that in the machine room is about 98–101
dB. The noise of the diesel locomotive is quite strong. The noise
level in the cab is about 99–108 dB and that in the machine room
is about 116–120 dB. The noise in the cab of a steam locomotive
is usually about 100 dB and about 125 dB in the machine room.
The comparison of noise generated by different locomotives is
shown in Figure 5.11.

(iii) *Aircraft noise*: Aircraft will also produce a lot of noise during
take-off, flight, landing, and ground tests. Aircraft noise mainly
includes propeller noise, exhaust noise, jet noise, fan noise, and
noise caused by boundary layer pressure fluctuation. The sound
made by the plane when taking off is about 140 dB. When an
ordinary civil aircraft flies smoothly, the noise near the wing
(engine) in economy class is 80–90 dB, the noise near the tail in
the back of economy class is 70–75 dB, and the noise in business
class is 70–80 dB. The noise comparison at different positions of
the aircraft is shown in Figure 5.12.

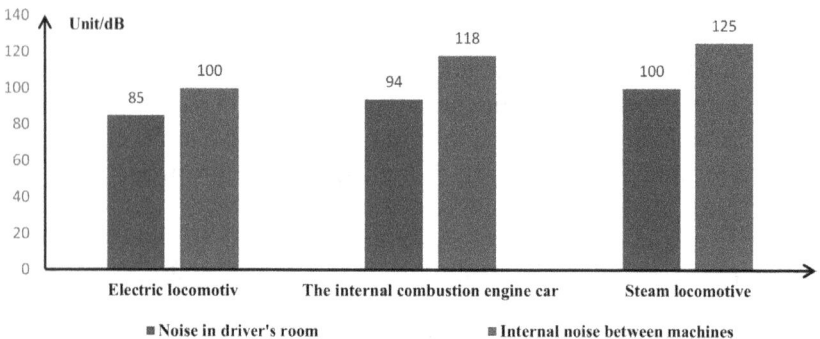

Figure 5.11   Comparison of noise generated by different locomotives.

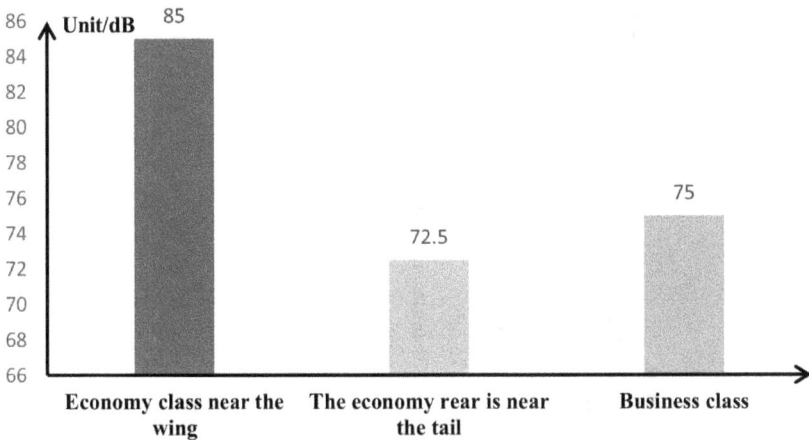

Figure 5.12 Noise comparison at different positions of the aircraft (unit: dB).

Table 5.7 Comparison of noise of various vehicles.

Vehicle	Ordinary railway	HSR	SSR	Automobile	Aviation
Noise per person km [dB/(km·person)]	0.1	0.05	0.01	1	1

(iv) *The noise of the SSR system*: The SSR system produces almost no external noise that affects people. This is because the SSR system does not contact the pipeline, so there is no vibration transmission. No noise from the capsule itself will be heard outside the tube, and the low air pressure in the tube can also prevent noise from spreading in the capsule. In the SSR system, the only potential noise source may be the vacuum pump. Of course, this noise is also very low. Table 5.7 shows the noise comparison of various vehicles.

In short, the SSR system shows absolute comfort compared with ships, cars, trains, and aircraft in terms of speed, cabin facilities, and noise impact. Taking the super train in the SSR system, people will experience a perfect journey. In addition, the SSR system also takes into account vibration, temperature, air, light, and other factors. The comfort comparison of various means of transportation is shown in Table 5.8.

Table 5.8 Comparison of comfort of various means of transport.

Vehicle		SSR	Ordinary railway	HSR	Automobile	Aviation
Inside the car stability	Longitudinal stability	1.5	3	2	3	3.2
	Lateral stability	0.2	2.2	2	2.5	2.6
	Vertical stability	1	2.5	2	2.8	5
Interior noise		50 dB	70 dB	65 dB	76 dB	80 dB
Interior temperature		Self-adjusting temperature	Higher than normal temperature	Self-adjusting temperature	Higher than normal temperature	Self-regulating normal temperature
Interior air		Second to outdoor	Same as outdoor	Second to outdoor	Same as outdoor	Second to outdoor
Interior light		Self-adjusting light	Same as outdoor	Second to outdoor	Same as outdoor	Self-adjusting light
Remarks		The smaller the stability value, the more stable and comfortable the interior environment is. The internationally recognized stability threshold is 2.				

5.2.3 *Stability of SSR*

Another major feature of the SSR system is the stability of its system. The SSR system mainly uses solar energy for power supply, which not only saves energy but also ensures the stability of the power supply system. The super train in the SSR system operates in the vacuum pipeline, and the propulsion system is built in the pipeline. The SSR system adopts a fully automatic control system for the operation process, due to which human errors are hard to avoid.

(1) **Stability of power system:** The power system is an important index to ensure train operation. The stability of the power supply system directly affects the safety and stability of train operation. The power source of the SSR system is mainly solar energy. The SSR system makes full use of the space above the pipeline and is covered with solar panels to convert solar energy into electricity. The energy obtained by the SSR system will exceed the energy consumed by the whole system. The SSR system also has energy storage facilities, which can operate the system for some time without using battery panels. On the one hand, the SSR track adopts maglev technology, which does not need a continuous power supply and greatly reduces the demand for energy consumption. On the other hand, the car body is suspended in a pipe close to a vacuum, and the resistance is very low. Therefore, the power consumption of the whole SSR system is low, and the operation can be maintained only by its solar power supply. In short, from the perspective of stability, compared with the HSR system, the power supply system of the SSR is not only more convenient and energy-saving but also more stable.

(2) **Stability of propulsion system:** The propulsion system of the SSR system is mainly air compression, as shown in Figure 5.14. The super train in the SSR system includes three stages: acceleration, constant speed, and deceleration to the station. The basic requirements for the propulsion system of SSR are as follows: to accelerate the cabin from 0 to 480 km/h, so that it can drive at a relatively low speed in the urban area and maintain the speed of 480 km/h if necessary. In the linear acceleration area, the operating speed can be accelerated from 480 to 1,220 km/h with an acceleration of 1 g (9.8 g/m^2). When approaching the station, the train can reduce the speed to 480 km/h. The fan at the head of the super train in the SSR system is sufficient

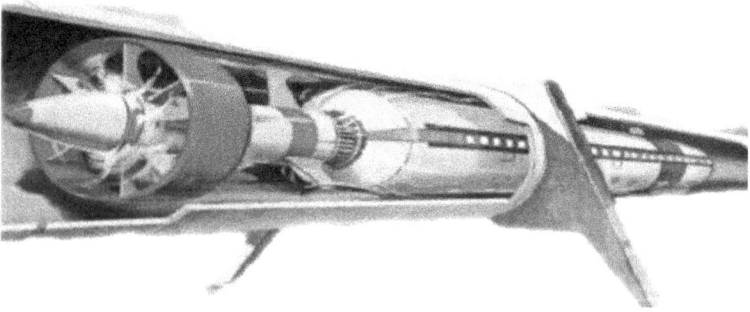

Figure 5.13   Propulsion system of SSR.

to provide the force for the train to maintain 1,220 km/h. The force during acceleration and deceleration is completed by the linear motor on the pipe wall, accelerating the train at departure and slowing it down as it nears a station. The propulsion system of the SSR is shown in Figure 5.13.

(3) **Stability of control system:** In the existing means of transportation (ships, cars, trains, and aircraft), full automation of the control system has not been realized, and most of them are controlled manually. During their operation, due to human errors and other factors, various accidents are easy to cause leading to serious consequences. Different from other means of transportation, the SSR system adopts a fully automated and information-based active control system. Through this automatic control system, the operation, control, and maintenance of the super train are automated, based on diagnostic technology, the complete informatization of operation and maintenance management is realized. Due to the automation and informatization characteristics of system technology, the stability of the SSR system is relatively high.

5.2.4 Reliability of SSR

Reliability is also an important technical index of vehicles. The SSR system also takes reliability into account in the design to ensure that passengers can be transported to their destination safely and reliably. The reliability of the SSR system is mainly reflected in the following:

(1) **Cabin reliability of SSR system:** The SSR system includes the super loop system with all infrastructure, mechanical, electrical, and software components in the engine room, which ensures its reliability, durability, and fault tolerance within its service life. The super train in the SSR system will not stop in the pipeline. If a train stops for some reason, the train in front of it will continue to go to the destination and will not be affected. The super train behind the stranded train will start the automatic emergency braking system and stop. Once the following super trains stop running, the stranded super train will use the small motor in the cabin to provide power and send itself to the destination safely.

(2) **Pipeline reliability of SSR system:** The SSR system operates in the pipeline, so its reliability is fully considered when constructing the pipeline. The pipeline of the SSR system is shown in Figure 3.4. It is envisaged that thick steel pipes will be used as the vacuum pipes. It is extremely difficult to pierce or break these pipes. While ensuring structural integrity, these pipes can resist pressure changes and gas leakage.

5.2.5 Economy of SSR

Economy is also one of the main conditions to be considered in the construction of the SSR system. The cost of the SSR system includes construction cost, operation cost, management cost, and maintenance cost.

Construction cost: The construction cost of the SSR system includes the cost of vehicles, pipelines, and stations.

Operation cost: The operation cost of the SSR system refers to the cost of the operation of vehicles and stations, such as energy consumption.

Management cost: The management cost of the SSR system includes real estate capital and employee cost.

Maintenance cost: The maintenance cost of the SSR system includes the maintenance cost of infrastructure and vehicles.

According to Elon Musk's assumption, two versions of the SSR system are being considered: the Passenger SSR system and the passenger-cargo SSR system. Due to their different requirements and functions, their costs are also different. The cost details of the SSR system as given as follows:

(1) **Construction cost of SSR system:** The construction cost of the SSR system mainly includes vehicle cost, pipeline cost, and station cost.

(i) *Vehicle cost of SSR system*: The vehicle cost of the SSR system includes the cost of the vehicle's external structure, interior decoration, battery, and propulsion system. The external structure of SSR includes the passenger compartment, window, door, and other structures. The overall cost of the external structure of the passenger SSR shall not exceed 245,000 dollars, and the overall cost of the external structure of the passenger and freight SSR shall not exceed 275,000 dollars. The internal structure of SSR includes seats, restraint systems, door panels, luggage compartment, and entertainment display. The total cost target of the overall internal components of the passenger SSR shall not exceed 255,000 dollars, and the total cost of the overall internal components of the passenger and freight SSR shall not exceed the 185,000 dollars. The expected vehicle cost of the SSR is shown in Table 5.9.

(ii) *Pipeline cost of SSR system*: Optimizing the inner diameter of the pipeline of the SSR system can alleviate the blocked airflow around the train and maintain a low material cost. To minimize the cost, an equal-thickness steel pipe is selected as the preferred material for the inner diameter steel pipe. The wall thickness of the passenger SSR pipeline is between 2.0 and 2.3 cm, including the prefabricated

Table 5.9   Expected vehicle cost of SSR.

No.	Vehicle components	Cost of passenger SSR (USD 10,000)	Cost of passenger and freight SSR (USD 10,000)
1	Overall architecture	24.5	27.5
2	Seats and interior trim	25.5	18.5
3	Propulsion system	7.5	8
4	Suspension and air bearing	20	26.5
5	Battery, motor, and coolant	15	20
6	Air compressor	27.5	30
7	Emergency braking	5	7
8	Management system	10	15
9	Total cost	135	152.5
10	Total of all trains	5,400	6,100

pipe section, reinforcing bar, and emergency exit. The cost of the pipe is expected to be less than $650 million. The pipe wall thickness of the passenger and freight SSR pipeline is between 2.3 and 2.6 cm. In this case, the cost of the pipe is expected to be less than $1.2 billion. To limit air leakage into the system, the vacuum pump will operate continuously at different positions along the pipeline to maintain the required pressure. The expected cost of all required vacuum pumps is not expected to exceed $10 million.

The pipes of the SSR system will be supported by pillars, and the construction materials will be reinforced concrete at a very low cost. The cost of the pillar structure and pipe joint of the passenger SSR pipeline is expected to be no more than $2.55 billion, and the cost of pillar structure and pipe joint of the passenger and freight SSR pipeline is expected to be no more than $3.15 billion. In addition, the super train needs solar panels to supply energy during its operation. The expected cost of the pipeline construction of the SSR is shown in Table 5.10.

(iii) *Station cost of SSR system*: The construction cost of the station of the SSR system mainly includes the external structure of the station and the internal infrastructure of the station. The design of SSR stations strives to be simple and practical. The boarding process and layout are much simpler than HSR stations and airports. The expected cost of each station is about $125 million. According to Musk's assumption, the full load of the train is 40 people and the total construction cost of the whole system of passenger SSR will be about $6 billion (Table 5.11).

Table 5.10   Expected cost of pipeline construction of SSR.

No.	Parts	Passenger SSR cost ($100 million)	Passenger and freight SSR cost ($100 million)
1	Pipeline structure	6.5	12
2	Vacuum pump	0.1	0.1
3	Tunnel engineering	6	7
4	Pipe support	25.5	31.5
5	Solar panels	2.1	2.5
6	Overall cost	40.6	50.6

Table 5.11   Total construction cost of the whole system of passenger SSR.

No.	Parts	Cost ($ million)
1	Engine room	54 (40 capsules)
2	Overall structure and door of engine room	9.8
3	Interior trim and seats	10.2
4	Compressor	11
5	Batteries and electronics	6
6	Propulsion system	5
7	Suspension and air bearing	8
8	Assembly cost	4
9	The conduit	5,410
10	Pipeline structure	650
11	Tower cost	2,550
12	Tunnel cost	600
13	Propulsion system	140
14	Solar panel	210
15	Station and vacuum pump	260
16	Land acquisition cost	1,000
17	Cost fluctuation	536
18	Total cost	6,000

The operation line of the SSR system is the pipeline, which is elevated and far from the ground, to reduce the occupation of land resources. The vacuum pipeline between two cities is built on the ground like the HSR. Where there are roads, there can be two pipelines for driving in two directions. Therefore, the construction cost of the SSR system is lower than that of other means of transportation. The comparison of construction costs between the SSR system and other modes of transportation is shown in Table 5.12.

(2) **Operation cost of SSR system:** The operation cost of the SSR system refers to the various costs and expenses incurred by the transportation enterprises in transportation production activities, such as employee wages, fuel, electricity, and depreciation of fixed assets. These expenses constitute the total operation cost. By comparing the

Table 5.12 Construction cost comparison between SSR and other modes of transportation.

Mode of transportation	Construction cost	Remarks
SSR	1	Set the cost of SSR as base 1
Freeway	4	
HSR	10	

operating costs of SSR, HSR, and aircraft, its economy is fully explained as follows:

(i) *SSR system operating costs*: The SSR system mainly uses solar energy for power supply, with a high energy utilization rate. In the process of operation, the full-automatic control system is adopted, which has little demand for employees. In addition, the design of the SSR ensures its service life, which greatly reduces its operation cost. Due to the reduction of contact friction and air friction, evacuated tube transport consumes less energy than any traditional means of transportation. The transportation capacity per kilowatt-hour of the SSR system is 50 times that of HSR. The SSR system will be powered by solar energy. After solar panels are installed in the system, the energy obtained will exceed the consumption. In addition, the SSR also has energy storage facilities, which can drive for a week without using battery panels.

(ii) *HSR system operating costs*: The operation cost of the HSR system mainly includes materials, fuel, electricity, and other expenses consumed by the transportation equipment during the operation of the HSR; salaries, bonuses, allowances, subsidies, welfare, and other expenses of personnel directly engaged in operation and production activities; depreciation of fixed assets; and seasonal loss during operation, shutdown loss during repair, and net accident loss. China's HSR with a speed of 350 km/h consumes more than 9,600 kW/h, and the HSR with a speed of 250 km/h consumes 4,800 kW/h. According to the electric charge of China's power grid, the industrial electricity charge is about 1 yuan/kWh, and the HSR with a speed of 350 km/h costs about 10,000 yuan/h. The depreciation of fixed assets of HSR mainly includes three parts: the depreciation of civil engineering, the depreciation of high-speed train vehicles, and the depreciation of traffic organization and dispatching communication signals and safety equipment. In general, the depreciation of fixed assets accounts for 45–50% of the total expenditure of HSR transportation and a high proportion of operating costs. The operation cost of the HSR is shown in Table 5.13.

(iii) *Aircraft operating costs*: The operating cost of aircraft mainly includes the following aspects: aircraft leasing fee; fuel cost during aircraft operation; labor cost, mainly including the salary and welfare cost of pilots, flight attendants, maintenance, and other related personnel; aircraft depreciation, mainly including amortization of aviation material consumables, amortization of high-priced revolving parts, and depreciation of aircraft during flight; the cost of meals, drinks, and other supplies is high on the plane; the take-off and landing fees charged by the airport company when the aircraft takes off and lands at the airport; and the shipping agency fee for outbound flights and the route fee levied by the control department. The biggest cost in aircraft operation is the fuel cost of the aircraft, which accounts for about 30–40% of the operating cost of the whole airline. A Boeing 737 with 160–189 passengers consumes about 2.5 to 3 tons of fuel per hour, and the cost per hour is between 11,250 and 13,500 yuan. The annual depreciation rate of the aircraft is about 10% of the present value, which is the main part of the aircraft operating costs.

Table 5.13   Operating costs of HSR system.

	Expenditure indicators	Cost
Energy consumption cost	Traction power supply fee (yuan/km)	0.02
	Other material expenditure (yuan/km)	0.0088
Employee salary	Salary of high-speed train attendants (yuan/km)	1.5
Station cost	Station service fee (yuan/thousand persons)	1,500
	Expenses related to station buildings (yuan/thousand people)	2,250
Depreciation expense	Depreciation of high-speed train (yuan/h)	2,480
	Depreciation of signal equipment (yuan/h)	580
Other expenses	Water supply station cost (yuan/km)	9
	Sewage station cost (yuan/km)	3.5

The operating costs of a certain type of aircraft are shown in Table 5.14.

In short, through the above analysis, the operation cost of the SSR system is shown to be much lower than that of the HSR and aircraft. The comparison of operation costs between the SSR system and other transportation modes is shown in Table 5.15.

(3) **Maintenance cost of SSR system:** All kinds of vehicles will have a certain degree of loss or even failure during operation, which requires daily maintenance or repair of vehicles, lines, and various components as follows:

(i) *Aircraft maintenance costs*: The maintenance cost of the aircraft includes the labor cost and material cost required for the airframe, engine, and components. It mainly includes route maintenance, system maintenance, engine maintenance, structural maintenance, and regional maintenance. Route maintenance is the lowest level and most basic maintenance activity, which can also be called daily maintenance and service. In terms of work content, the main cost of route maintenance is labor cost, accounting for about 13% of the total maintenance cost. In

Table 5.14   Aircraft operating costs.

No.	Cost index	Cost (yuan/h)
1	Fuel cost	2,802
2	Unit cost	300
3	Crew costs	140
4	Landing fee	2,600
5	Catering expenses	425
6	Insurance premium	636
7	Depreciation charge	4,453

Table 5.15   Comparison of operating costs between SSR and other modes of transportation.

Mode of transportation	Operating costs	Remarks
SSR	1	Set the cost of SSR as base 1
HSR	3	
Aircraft	6	

addition, letter inspection (A–C inspection) of the aircraft accounts for a small proportion of the total maintenance cost, accounting for about 7%. Aircraft overhaul (D inspection) accounts for about 13%. Component maintenance accounts for about 28% of the total maintenance cost. The engine is one of the most important parts of the aircraft, so its repair accounts for the largest proportion, about 39%. The distribution of aircraft maintenance costs is shown in Figure 5.14.

(ii) *HSR maintenance costs*: The maintenance of the HSR mainly refers to the labor, materials, and other costs consumed in the maintenance and repair of transportation equipment during the production process. It includes regular maintenance cost of the HSR, HSR overhaul cost, communication signal maintenance cost, line maintenance cost, and traction power supply equipment maintenance cost. The maintenance cost of HSR is shown in Table 5.16.

(iii) *SSR system maintenance costs*: The SSR system has a high speed and high utilization rate, resulting in relatively high maintenance costs. However, the maintenance cost of pipelines, stations, and rolling stock in the SSR system accounts for less than 30% of the infrastructure cost, and the ratio of maintenance cost to construction cost is less than that of the railway. From this point of view, the relative maintenance cost of the SSR system is not high.

(4) **SSR system management costs:** The management cost of the SSR system is mainly the cost of management personnel. As for the cost of management personnel, because the whole SSR system is fully

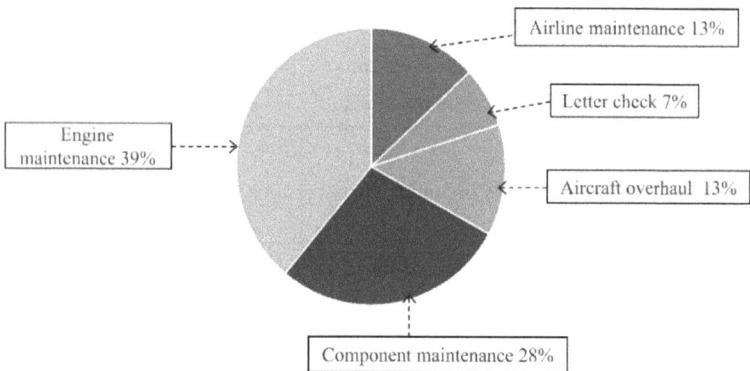

Figure 5.14 Aircraft maintenance cost distribution.

Table 5.16   Maintenance cost of HSR.

No.	Cost index	Cost
1	Regular maintenance cost of high-speed train (yuan/km)	26
2	Overhaul cost of high-speed train (yuan/km)	10
3	Communication signal maintenance cost (yuan/km)	22
4	Line maintenance cost (yuan/ton km)	0.0102
5	Maintenance cost of traction electric equipment (yuan/ton km)	0.0135

Table 5.17   Factors affecting transportation costs.

Influence factor	Cause of influence	Form of expression
Transport distance	It directly affects variable costs such as labor, fuel, and maintenance	The allocation of fixed costs decreases with the increase of transportation distance
Carrying capacity	Effect on variable costs	The allocation of fixed costs decreases with the increase of freight capacity
Application of loading capacity of means of transport	Transportation vehicles are more limited by volume, which affects variable costs	The more fully utilized the loading capacity, the lower the transportation cost
Transportation accident loss	Impact indirect costs	The loss caused by the accident increases the indirect cost
Imbalance of transportation demand	Affect the allocation of transportation capacity and the economy of transportation	Direction imbalance will cause idling, and time imbalance will cause the loss of opportunity cost and fixed cost

automated, there are not many personnel needed. The cost of management personnel is lower than that of operators. After these costs are included, the cost of operation personnel increases by 10%. The management expenses of the SSR system mainly include the salaries, bonuses, allowances, and subsidies of enterprise managers. This part of the cost is generally calculated in the operating cost, which is close to the cost of operators. The operation cost, maintenance cost, and management cost of transportation tools can be collectively referred to as transportation cost. Due to the different characteristics of various transportation tools, the transportation costs are also different. The factors affecting transportation costs are shown in Table 5.17.

Table 5.18 Advantages and disadvantages of transportation costs of various means of transportation.

Vehicle	Advantage	Disadvantage
Road transport	Less construction investment, fast capital turnover, and short payback period	Energy consumption and unit transportation cost are high, which is not suitable for bulk and long-distance transportation
Waterway transportation	Low transportation cost and less investment	Unable to form a national water transportation network
Railway transportation	No raw material expenditure and low transportation cost	Large investment in infrastructure construction
Transport aviation	Short construction period, less investment, and fast investment recovery	The aircraft cost is very high, the transportation cost is high, the energy consumption is large, and the relative transportation capacity is small
Pipeline transportation	Low energy consumption and low transportation cost	The investment in infrastructure construction is large, and the consumption of metals is also large

The characteristics of transportation costs of various means of transportation are also different. According to the analysis of transportation cost of various means, the advantages and disadvantages of various means of transportation are shown in Table 5.18.

5.3 Summary

As a new mode of transportation, the SSR system has absolute advantages both from the perspective of systems science and economic cost. Compared with other transportation modes, the SSR system has the best speed, strong transportation capacity, good safety, low energy consumption, little impact on the environment, comfort and convenience, and considerable economic benefits. With its unique technical advantages, it meets the increasing demand for modes of travel. Once put into operation, it will be a milestone in long-distance transportation. Many characteristics of the SSR system may trigger a revolution in the field of transportation and promote the progress of human society in the future. However, there are still many problems at the technical level, which require scholars to continue to study and explore a solution step by step in a sustainable manner.

Chapter 6

Development Trend
of Super-Speed Rail (SSR)

The SSR system is a long-distance transportation mode, which can allow people to travel at the speed of 1,200 km/h. In essence, the SSR system is a long pipe that discharges air to form a vacuum space. It is suspended on the ground to reduce the impact of weather and natural phenomena like earthquake. Passengers can choose a single cabin or multi-person cabin, and then the SSR accelerates under the action of magnetic force. So, compared with other means of transportation, is speed the only advantage of the SSR system?

6.1 Advantages of SSR

According to Elon Musk's assumption, the expected speed of the SSR system is 1,200 km/h, close to the speed of sound at 340 m/s. This speed will be two or three times faster than the fastest bullet train now and twice as fast as a plane. The super train takes 28 people. The fare from Los Angeles to San Francisco is $20. It can transport 7.4 million passengers a year and recover the investment in 20 years. At present, the United States, China, and Europe are carrying out tests and preparations for the SSR.

6.1.1 *Speed*

Speed has always been the eternal pursuit of mankind. Since the advent of the steam train in the 19th century, human beings have constantly transformed the means of transportation. From bicycles, motorcycles, cars, planes, trains, to even maglev trains, the speed of vehicles is increasing more and more, as we also witness the rapid development of science and technology and constantly refresh people's understanding of speed. Therefore, the speed of operation is one of the criteria for measuring the advanced nature of transportation.

(1) **The speed of the railway system:** The current railway system is divided into ordinary-speed railway system, rapid railway system, and high-speed railway system:

(i) *Ordinary-speed railway system*: Ordinary-speed railway is called ordinary rail for short. Ordinary-speed railway system refers to the railway with low design speed and can only allow trains to run at ordinary speed. According to China's standard for ordinary speed railway, this refers to dedicated non-passenger lines with a speed of no more than 160 km/h and dedicated passenger lines with a speed of no more than 140 km/h. The most common ordinary-speed train is known as the "green train," as shown in Figure 6.1. After the speed of trains increased, the concept of grade based on different speed indicators as the basis for railway division became more and more popular. Although the speed grade of ordinary-speed railway is low, it is the most common in the world railway system.

(ii) *The speed of the rapid railway system*: The rapid railway system refers to the railway whose design speed standard is between the

Figure 6.1   Normal-speed train.

Figure 6.2 Fast rail train.

ordinary-speed railway and high-speed railway. The different speeds of railways have long been distinguished internationally. China has introduced the concept of the fast railway to better classify railways according to speed grade. Rapid rail, as shown in Figure 6.2, is the mainstream of China's modern railway construction. In the long-distance field, it is a mixed passenger and freight line to meet a variety of needs. In China, in the era of high-speed railway, rapid railway refers to the railway with a design speed of 160–200 km/h (including reserved).

(iii) *Speed of high-speed railway system*: The high-speed railway is referred to as the HSR for short. An HSR system refers to the railway system with a high infrastructure design speed standard that can provide safe and high-speed running of trains on the track. HSR has different regulations in different countries and different durations, based on different scientific and academic studies. The State Railway Administration of China defines China's HSR (Figure 6.3) as a new or existing dedicated passenger train with a designed operating speed of more than 250 km/h. China has promulgated the corresponding medium and long-term railway network planning documents, which include some 200 km/h track lines into the scope of China's high-speed railway network.

(2) **The speed of maglev traffic:** Based on the principle of "opposites attract and likes repel," a maglev train is a vehicle that uses

Figure 6.3   High-speed train.

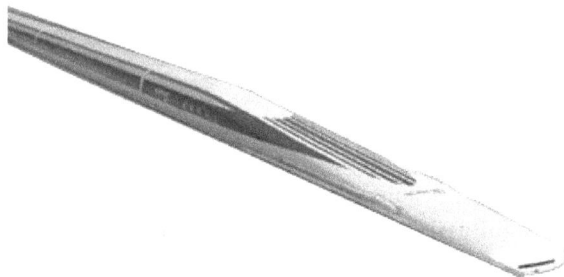

Figure 6.4   Maglev train.

electromagnetic force to suspend the vehicle body on the track and drive the vehicle through electromagnetic force. The train does not have contact with the ground during operation, which eliminates the friction resistance of the wheel-rail system and can reach a very high running speed. Therefore, the maglev train is a modern high-tech rail transit. The maglev train is shown in Figure 6.4.

The biggest feature of the maglev train is no friction resistance during operation, with only air resistance left, and a running speed that can exceed 400 km/h, or even more than 600 km/h. For example, the running speed of ultra-high-speed maglev trains in Germany exceeds 430 km/h, while the highest speed record of the ultra-high-speed maglev train in Japan is 603 km/h. Because the maglev train is not affected by friction resistance, it can reach a very high speed in theory. However, due to the limitation of air resistance, it can only run at 400–1,000 km/h. The operation speed of 1,000 km/h is the early

Table 6.1 Types of maglev trains.

No.	Types	Speed (km/h)	Name	Major countries
1	First type	400–600	Low-temperature maglev train	Japan, Germany
2	Second type	600–800	Normal-temperature maglev train	Japan
3	Third type	800–1,000	High-temperature maglev train	—

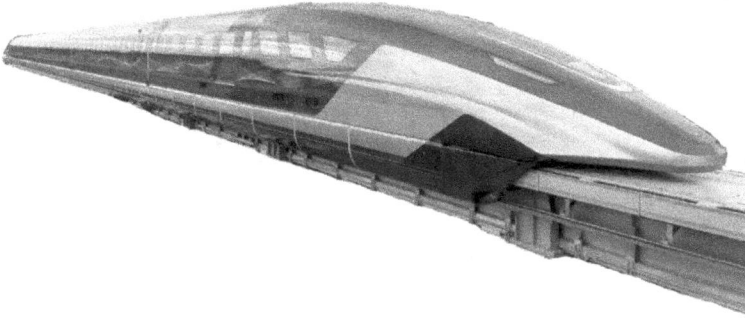

Figure 6.5 Maglev train in China.

warning threshold of a maglev train, and the operation cost will be too high if it exceeds this speed. In terms of suspension mechanism, the maglev train can be divided into electromagnetic suspension (EMS) and electrodynamic suspension (EDS). In this book, maglev trains are divided into three types, as shown in Table 6.1.

The maglev train is a train that uses superconducting magnets to float the car body and obtain propulsion power by periodically changing the direction of magnetic poles. In addition to high speed, the maglev train also has the characteristics of no noise, no vibration, and energy saving. It is expected to become the main means of transportation in the 21st century. A maglev train in China is shown in Figure 6.5.

(3) **Speed of passenger aircraft system:** The passenger aircraft is a collective flying means of transport of a large size and large passenger capacity, which is used for domestic and international commercial travel. Commercial airliners are mainly divided into trunk airliners and branch airliners. Airliners are divided into short-range, medium-range, and long-range airliners according to their range. They are also

Figure 6.6   Passenger aircraft (Boeing 787).

Figure 6.7   C919.

divided into small, medium, and large according to take-off weight and passenger capacity. There are also propeller airliners and jet airliners based on the driving mode. A passenger plane is shown in Figure 6.6.

The speed of different types of passenger aircraft is different, generally about 700–1,000 km/h. The cruise speed of Boeing 737 can reach Mach 0.75, that is, nearly 918 km/h, and the cruise speed of Boeing 747 can reach Mach 0.98, nearly 1,120 km/h. As the fastest means of transportation at present, the aircraft reduces the time of long-distance travel greatly, and the passenger capacity is also considerable. It is the first choice for people to travel across borders. An airline is shown in Figure 6.7.

(4) **The speed of the SSR system:** Because the maglev train is not hindered by friction resistance but is affected by air resistance, the running speed will have a limit. To reduce the air resistance, further

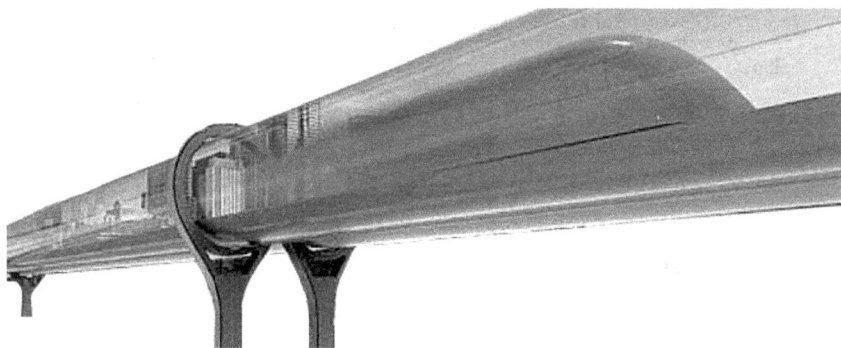

Figure 6.8   SSR.

Table 6.2   Types of SSR.

No.	Types	Speed (km/h)	Name	Major countries	Remarks
1	First type	1,000–1,200	Low-speed SSR	USA	Sound speed:
2	Second type	1,200–10,000	Medium-speed SSR	NA	340 m/s, that
3	Third type	10,000	Hypersonic SSR	NA	is, 1,224 km/h

improve the operation speed, and meet the requirements of human rapid travel, based on the concept of "vacuum pipeline," the SSR is proposed, and the operation speed is expected to be more than 1,200 km/h. The design of the SSR is shown in Figure 6.8.

Because there is no friction resistance and air resistance in the vacuum pipe, the super train can run "willfully," and the theoretical maximum speed can reach more than 10,000 km/h. The super train may become a new generation of transportation after automobiles, ships, trains, and aircraft. This book divides the SSR system into three types, as shown in Table 6.2.

The common means of transportation are automobiles, trains, ships, and aircraft. A specific comparison of the operation speed of various main means of transportation is shown in Table 6.3 and Figure 6.9.

6.1.2 *Energy consumption*

The transportation industry is a large energy consumer. General cars use gasoline or kerosene as liquid fuel, and the engine generates heat to drive

Table 6.3   Speed of different travel tools.

No.	Name	Speed (km/h)	Remarks
1	Automobile	40–120	Sound speed: 340 m/s,
2	Ordinary-speed railway	<160	that is, 1,224 km/h
3	Express Railway	160–250	
4	HSR	>250	
5	Magnetic suspension	400–1,000	
6	Passenger aircraft	700–1,000	
7	SSR	>1,000	

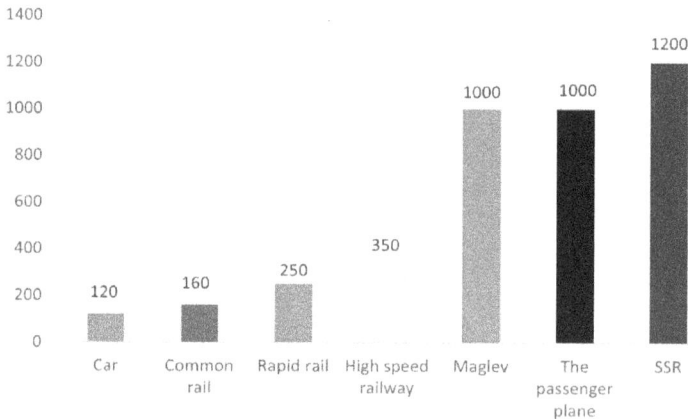

Figure 6.9   Speed of different travel tools (unit: km/h).

the car. The train uses electric traction, does not consume valuable liquid dyes such as oil, and can use various forms of energy.

Laying solar panels on the pipes of the SSR system can generate enough electric energy to maintain low-energy operation (about 1/3 of the aircraft energy consumption), and even enable the SSR to store energy. The average energy consumption per kilometer of various means of transportation is shown in Table 6.4 and Figure 6.10.

6.1.3 *Transportation capacity*

The SSR system has a large transportation capacity. According to Elon Musk's plan, the SSR can transport 164,000 passengers in one direction

Table 6.4 Average energy consumption per kilometer of different transportation modes.

Mode of transportation	Ordinary train	HSR	Bus	Car	Aircraft	SSR
Energy consumption [g/(person·km)]	403.2	571.2	583.8	3,309.6	2,998.8	999.5
Energy	Power	Power	Gasoline or kerosene	Gasoline or kerosene	Gasoline or kerosene	Power

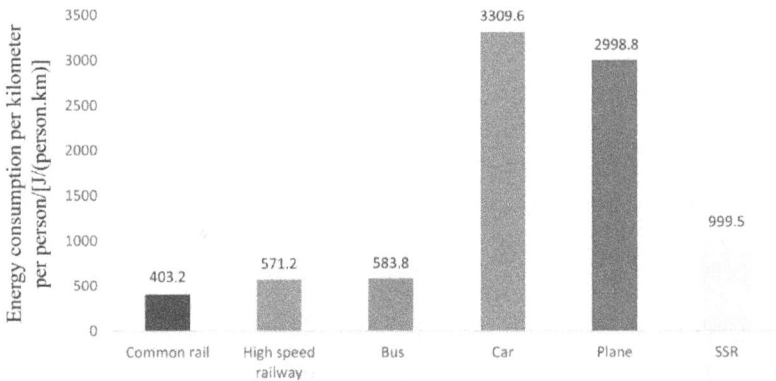

Figure 6.10 Average energy consumption per person km of different transportation modes [unit: g/(person·km)].

every day and start every 40 s, that is, 90 trains can start every hour and 59.86 million passengers a year. The maximum passenger volume of an expressway in a year will not exceed 10 million person times. An HSR can carry 150 million passengers a year. According to the data, the transportation capacity of the SSR system is less than that of an HSR system. For example, the carrying capacity of HSR is about 10 times that of air, 5 times that of the expressway, and 2.5 times that of SSR, while the carrying capacity of SSR is about 3 times that of air and 1.5 times that of the expressway. However, the transportation cost of Evacuated Tube Transport will be very cheap, only 1/4 of Expressway and 1/2 of the HSR. The transportation cost of HSR is 1/5 of that of air and 2/5 of that of the expressway. The comparison of transportation capacities is shown in Table 6.5, and the maximum transportation capacity of different modes is shown in Figure 6.11.

Table 6.5   Transportation capacity of different transportation modes.

No.	Type of shipping	SSR	HSR	Expressway	Transport aviation
1	Carrying capacity (based on HSR)	1	2.5	1/1.5	1/4
2	Transportation cost (based on HSR)	1	2	4	10

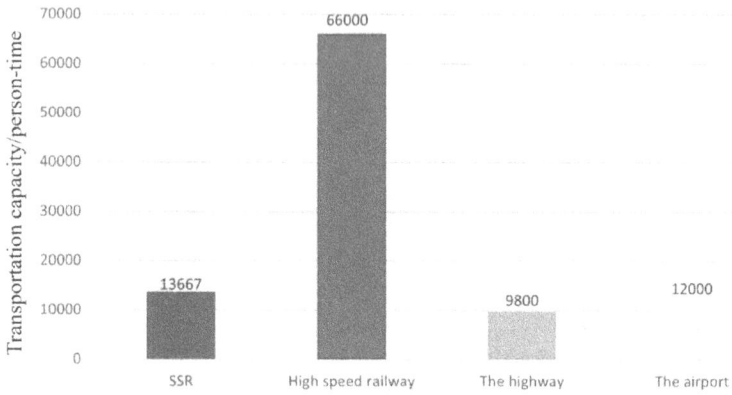

Figure 6.11   Comparison of hourly two-way maximum transportation capacity between SSR and other main transportation modes (unit: person time).

6.1.4 *Safety*

Since the advent of the HSR, Japan, Germany, France, and China have transported 10 billion passengers. No major traffic accidents have occurred in HSR in the various countries, and no casualties have been caused by accidents. This is rare in all kinds of modern transportation modes. In particular, HSR countries send thousands of high-speed trains a day. Even if five accidents occur, the accident rate and casualty rate are far lower than other modern transportation modes. Therefore, HSR is considered to be the safest means of transportation. Statistics show that the world's road accidents generally kill 300,000–500,000 people every year, and the average number of deaths per 1 billion person kilometers is as high as 140. Every year, about 50 planes crash in global civil aviation accidents, killing more than 2,000 people. Therefore, the number of deaths per 10 person kilometers is 1.971 for railways, 18.929 for cars, and 16.006 for planes. Table 6.6 shows the number of deaths per 1 billion person kilometers in Japan in 1985.

Table 6.6 Deaths per 1 billion person kilometers in Japan (1985).

No.	Type of shipping	Death toll
1	Railway	1.971
2	Automobile	18.929
3	Aircraft	16.606

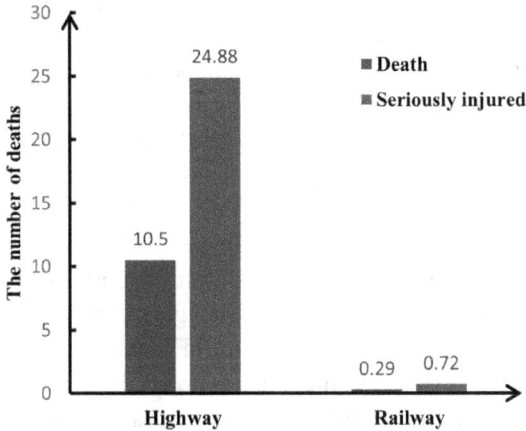

Figure 6.12 Number of deaths and injuries in traffic accidents per 100 million person kilometers of roads and railways in China (unit: person).

According to the data released by the China Academy of Railway Sciences in research on "social cost of China's HSR and its contribution to society," among the traffic accident deaths and injuries per 100 million person kilometers in China, 10.5 people died on the highway (24.88 seriously injured) and 0.29 people died on the Railway (0.72 seriously injured), as shown in Figure 6.12.

For various modes of transportation in the atmospheric environment, including aviation, shipping, highway, and railway, weather and environment are often decisive factors. However, for evacuated tube transport, the impact of weather and environment is fundamentally limited. Therefore, generally, people do not need to consider the impact of wind, fog, rain, and snow on vacuum pipeline traffic. Although the SSR system has not been put into commercial operation and there are no detailed accident statistics, it is generally believed that the safety of a vacuum tube SSR system is higher than that of an HSR system from the perspective of operation technology and operation environment.

6.1.5 *Comfort*

The HSR has large vehicle space, and passengers are more comfortable lying, sitting, and traveling in it than other modes of transportation. The high-speed train has a luxurious interior layout, complete working and living facilities, spacious and comfortable seats, good walking ability, very stable operation, shock absorption, sound insulation, and quiet interiors. In addition to the advantages of the high-speed train, the super train is specially designed to take into account the safety and comfort of passengers, and the seats are well in line with ergonomics. Although the vacuum transportation pipeline can reach an incredible speed, passengers can only feel a small explosive acceleration force, thus maintaining comfort and safety during high-speed acceleration. In addition, because the transport cabin of the SSR system operates in the pipeline, passengers cannot see the scenery outside the window with their own eyes. Considering this, augmented reality windows will be used in the transport cabin, which can not only show the real world outside the window but also add digital information with entertainment attributes so that passengers can fully enjoy the journey. Some comfort indicators are shown in Figure 6.13.

When the train is running, there are certain noise problems. The international standard for environmental protection is that the noise level measured at a place 25 m away from the vehicle center and 1 m above shall not exceed 75 dB. When the train exceeds 200 km/h, the noise may exceed this standard. Excessive noise is often an important factor in speed

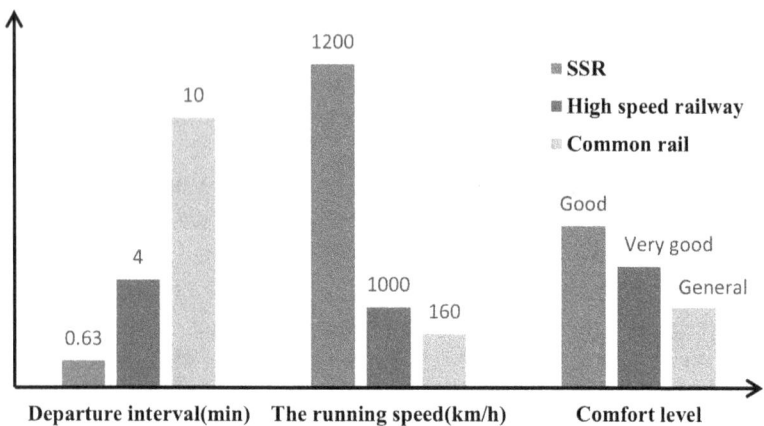

Figure 6.13   Comfort comparison between SSR and wheel-rail railway.

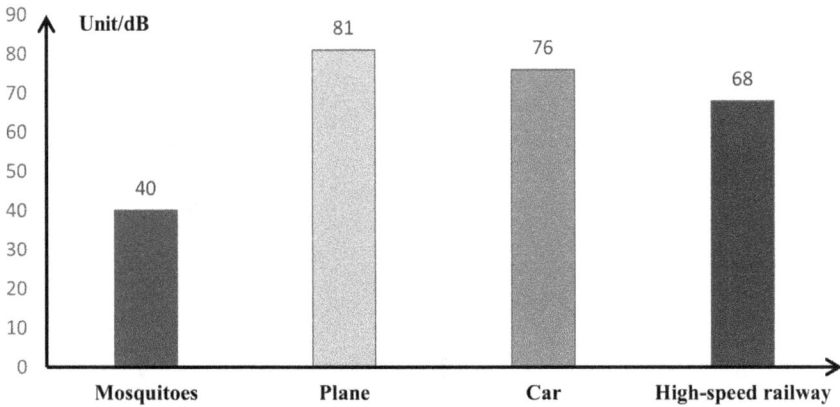

Figure 6.14   Comparison of noise of different vehicles (unit: dB).

limitation. For example, the noise of the maglev train in Germany reaches 89 dB when the speed is 400 km/h. When the speed is 430 km/h, the noise reaches 93 dB. It is generally believed that the maglev train has no noise and vibration without wheel-rail contact, which can only be possible at low speed because it is dominated by mechanical vibration and noise. As soon as the speed is increased, it will be dominated by aerodynamic noise, and its intensity will increase sharply with the sixth to eighth power of the speed. This is an objective law for any vehicle in the atmosphere. Whether it is suspended or not makes no substantive difference. It has a non-negligible impact on the noise pollution along the track and hurts people's psychology and physiology.

In the SSR system, the train running in the vacuum pipe will not have pneumatic noise. Statistics show that the sound of a mosquito is 40 dB, the noise in an aircraft cabin during flight is about 81 dB, the noise in a car at the speed of 120 km/h is about 76 dB, the noise in a carriage of China's HSR EMU at the speed of 380 km/h is about 68 dB, and the noise in a carriage of the SSR will be about 0 due to its operation in a vacuum tube. The specific data are shown in Figure 6.14.

6.1.6 *Economy*

The construction cost needs to be considered in the construction of transportation. If the construction cost is too high, resulting in high operation

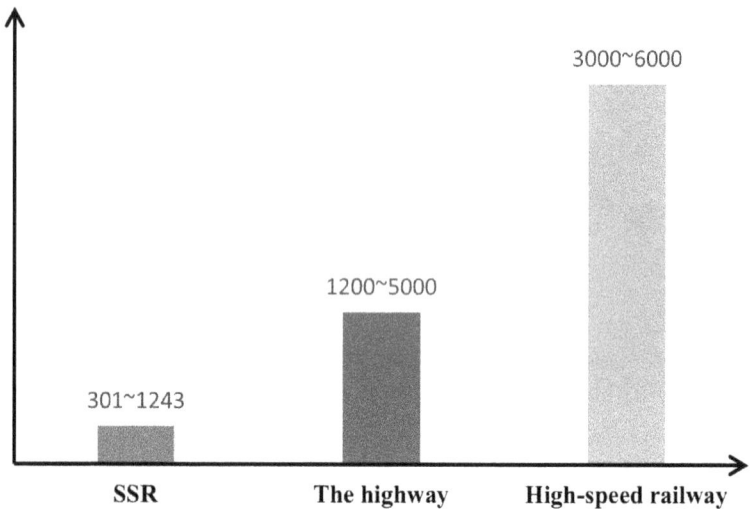

Figure 6.15   Construction cost of SSR and other modes of transportation (unit: US $10,000).

costs, it is bound to increase the travel cost for passengers. If passengers do not choose this mode of travel because of economic pressure, it will also increase the financial burden for each operating company. The construction cost of an HSR is about $30–$60 million per kilometer (China's HSR construction cost: the construction cost of 250 km/h HSR is 80 million yuan/km, and the construction cost of 350 km/h HSR is about 130 million yuan/km). As for the construction cost of the SSR system, Elon Musk estimates that the cost from San Francisco to Los Angeles will be $6 billion, or about $11.5 million per mile. Hyperloop transportation technologies announced that they could control the cost per mile at $5–$20 million, about $3.01–$12.43 million per kilometer. Because the SSR system uses a sealed vacuum pipeline to replace the track, the cost has been greatly reduced, and only 1/10–1/5 of the HSR cost is required. Figure 6.15 shows the specific comparison of transportation costs.

6.2 Development Vision of SSR

Among the seven continents in the world, except for Antarctica, each continent has countries. The HSR system can connect different regions and promote regional integration. Different types of HSR promote different types of regional integration. For example, the wheel-rail HSR system promotes regional integration. The maglev HSR system promotes the integration of all continents (Asia–Europe integration, Europe–Africa integration). The SSR system promotes global integration, that is, the global village. Moreover, after the opening of the SSR system, it will not only be a transportation channel connecting China and surrounding regions but also a starting point for the construction of railway networks connecting China and other countries on all continents. It will promote the connection between China and all countries, which is of great significance to regional cooperation and intercontinental integration.

6.2.1 *Integration of all continents*

If the world's continents can be connected by the HSR continental bridge, according to the existing HSR operation speed, it will take less than two days from East Asia to North America. However, for the SSR system, it may take only 2 h to realize this distance. Therefore, for the SSR system, it is extremely easy for all countries in the world to achieve integration in the following manner:

(1) **European integration based on SSR system:** Europe is located in the northwest of the eastern hemisphere, bordered by the Arctic Ocean in the north, the Atlantic Ocean in the West, and the Mediterranean Sea and the Black Sea in the south. The continent extends to the Polar Ural Mountains in the east, Cape Trypiti in the south, Cape Roca in the west, and North Cape in the north. Europe is bounded by the Ural Mountains and the Ural River in the east; the Caspian Sea, the great Caucasus mountains and the Black Sea in the southeast; North America across the Atlantic Ocean, the Greenland Sea, and the Danish Strait in the west; the Arctic Sea in the north; and Africa across the Mediterranean Sea in the south (the dividing line is the Strait of Gibraltar). Europe is the second smallest continent in the world, only a little larger than Oceania. It is called the Eurasian continent together with Asia and Eurasian African continent together with Asia and Africa.

Europe has developed HSR lines. And formed the European HSR network in 2020, but the research and development of SSR was lagging behind. However, the Hyperloop One will try to enter Finland and Sweden: According to the Daily Mail, the world's first full-scale

"SSR" in Europe will connect Helsinki, the capital of Finland, and Stockholm, the capital of Sweden, with a speed of 700 mi/h (about 1,126 km/h). Helsinki and Stockholm are 310 mi apart. It takes one hour by plane and 16 h by boat. After the completion of the SSR system, the round-trip time can be reduced to 28 min. Elon Musk said that the SSR is only suitable for cities with a distance of less than 1,609 km. As the SSR is still in the experimental stage and has not been truly commercialized, there is still some time before the real construction of SSR systems in various countries commences. However, we can be sure that the SSR network of each continent will be established first among important cities of each country (such as the capital). The schematic diagram of the SSR network in Europe is shown in Figure 6.16.

Figure 6.16   Schematic diagram of European SSR network.

(2) **Asian integration based on SSR system:** Asia is the largest and most populous of the seven continents. It covers 8.7% of the total area of the earth (or 29.4% of the total land area). Most of Asia is located in the northern and eastern hemispheres. The dividing line between Asia and Africa is the Suez Canal. East of the Suez Canal is Asia. The dividing line between Asia and Europe is the Ural Mountains, the Ural River, the Caspian Sea, the great Caucasus Mountains, the Turkish Strait, and the Black Sea. Asia lies to the east of the Ural Mountains and south of the Great Caucasus Mountains, the Caspian Sea, and the Black Sea. The west is connected with Europe, forming the largest landmass on earth, Eurasia.

Trans Asian Railway (TAR) has been planned in Asia, which is a unified freight railway network connecting Eurasia. Representatives of 18 Asian countries officially signed the intergovernmental agreement on the Asian Railway Network in Busan, South Korea, on November 10, 2006. The Pan Asian railway network proposal, which has been planned for nearly 50 years, was finally implemented. According to the agreement, the Golden Corridor composed of four "steel silk roads" can connect the two continents of Europe and Asia, and the crisscross trunk lines and branches will weave a huge economic cooperation network. However, the SSR system is different from the HSR. Due to its operating speed and pipeline laying, it is closely dependent on the environment, population, economy, and terrain of the regions. The schematic diagram of the Asian SSR network is shown in Figure 6.17.

(3) **American integration based on SSR system:** America is the combined name of South America and North America. It is also the abbreviation of "Americana," also known as the new continent. Northern Asia and the Americas is a low-lying region in the northern hemisphere. It is bounded by the Atlantic Ocean in the East, the Pacific Ocean in the west, the Arctic Ocean in the north, and the Panama Canal in the south. North America includes not only the Americas north of the Panama Canal but also the West Indies in the Caribbean Sea. South Asia and the Americas is a low-lying region. South America is located in the south of the Western Hemisphere, bordered by the Atlantic Ocean in the East, the Pacific Ocean in the west, the Caribbean Sea in the north, and Antarctica across the Drake Strait in the south.

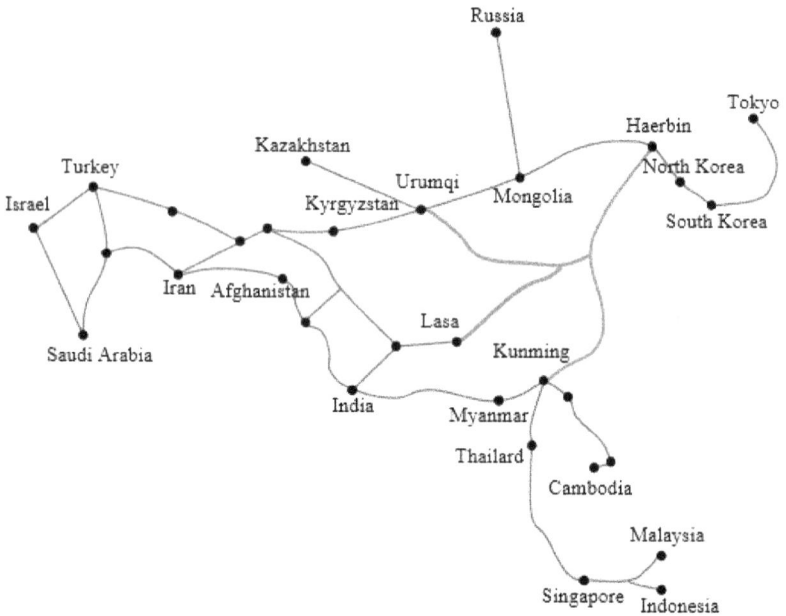

Figure 6.17   Schematic diagram of Asian SSR network.

At present, except for the United States, which has the SSR test plan from Los Angeles to San Francisco, California, other countries in the Americas have not yet planned the SSR. Therefore, the American Integration of SSR refers to the planning of the HSR network in America. The SSR network plan in North America and South America is shown in Figure 6.18.

(4) **African integration based on the SSR system:** Africa is located in the westernmost part of the eastern hemisphere and is the second-largest continent in the world. Africa faces the Atlantic Ocean and the Indian Ocean in the east, Europe across the Mediterranean in the north, Asia in the northeast, and the equator across the middle. There are 57 countries and regions, of which Sudan has the largest area and Nigeria has the largest population.

An HSR system is a project with high input and high cost. At present, lines with good comprehensive benefits of HSR are generally distributed among urban agglomerations with high population density in developed economies all over the world. Even as the population of large cities in Africa has expanded rapidly in recent decades, the

Figure 6.18 Planning of SSR network in America.

degree of economic development has not improved accordingly. It is difficult to support the cost of HSR operation in the short term. Therefore, when Africa has not established a mature HSR network, an SSR system plan cannot be ideated. Therefore, the integration plan under the African super railway system can refer to the African HSR network plan, which is mainly to be built between the major cities along the line as shown in Figure 6.19.

6.2.2 *Regional integration*

The HSR has created new growth points for urban development, promoted the integration of central cities and satellite towns, strengthen the radiation effect of the central city on the surrounding cities, and strengthen the "one-city effect" of the adjacent big cities. Especially with the vigorous development of HSR around the world, in the foreseeable future, HSR will become the main ground transportation mode connecting

Figure 6.19   SSR network planning in Africa.

countries in various regions, opening the era of the "global village" together with air routes, and travel around the world by HSR will become a tourism attraction. Regional integration under the SSR system is shown in Figure 6.20.

(1) **Asia–Europe integration based on SSR system:** The SSR system is a new mode of transportation after HSR. With the continuous improvement of technology, it is more and more possible to build the SSR system among countries and regions all over the world. Due to the characteristics of high safety, many trains, and short interval of the SSR system, it will be easy to realize the "one-city effect" in the world. The Asia–Europe integration plan under the SSR system is shown in Figure 6.21.

(2) **European–African integration based on the SSR system:** For the European–African integration under the SSR system, the Spanish railway can be considered to connect with Africa through the Strait of

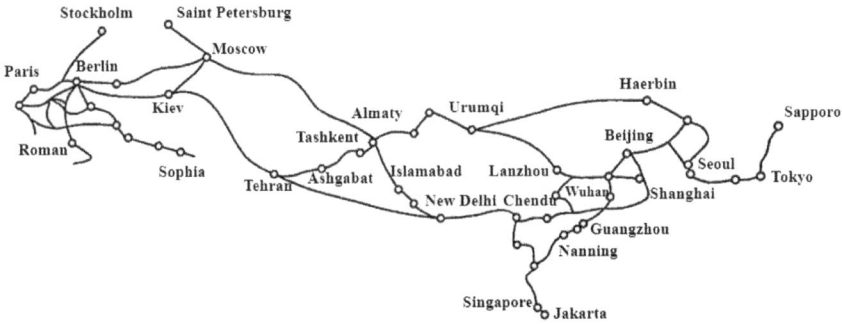

Figure 6.20 Regional integration under SSR system.

Figure 6.21 Asia–Europe integration plan under SSR environment.

Gibraltar subsea tunnel. Therefore, the European–African SSR system diagram based on the HSR network is shown in Figure 6.22.

6.2.3 Global integration

Compared with the subsequent development of other transportation modes, the rail transit system is slower than aviation, less numerous than cargo ships, and less flexible than cars, which makes the development of

Figure 6.22   European–African integration in SSR environment.

railway transportation relatively stagnant. However, due to the unique-ness of the rail transit system, especially the maturity of HSR technology in recent decades, the advantages of rail transit systems have become more prominent. Global integration in the HSR environment is shown in Figure 6.23.

At present, the actual operating speed of HSR in the world is mostly below 350 km/h. Compared with aviation, this speed is not enough to support an international transportation system (Figure 6.23). However, research on new HSR in countries such as China, Japan, Germany, France, and the United States is still very active. In the foreseeable future, with the improvement of technology and the vigorous promotion of the SSR system, the SSR system will approach global integration as shown in Figure 6.24.

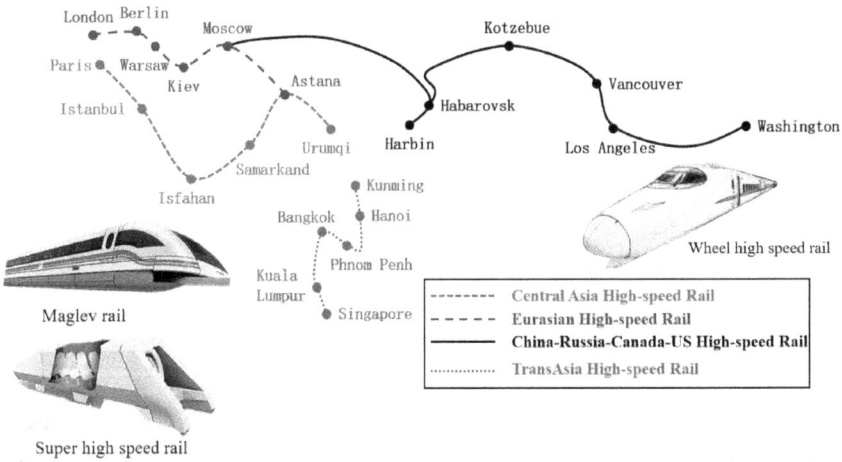

Figure 6.23   Global integration in HSR environment.

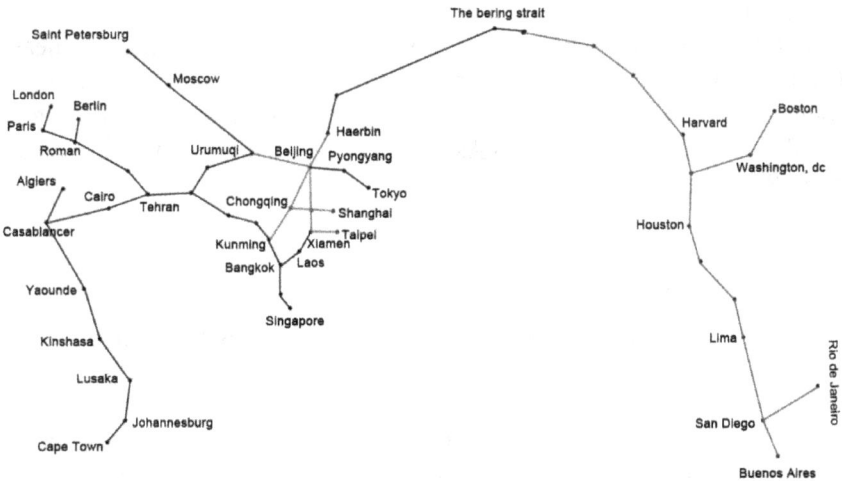

Figure 6.24   "Global village" under SSR environment.

Elon Musk said that the cost of building the "Hyperloop" transportation system from Los Angeles to San Francisco will only be $60 billion if the system only carries passengers and $75 billion if the system wants to carry people, goods, and vehicles. After calculating the ticket price of advanced SSR, the cost of building the SSR between Los Angeles and San Francisco will be only one-tenth that of an ordinary HSR. The two cities

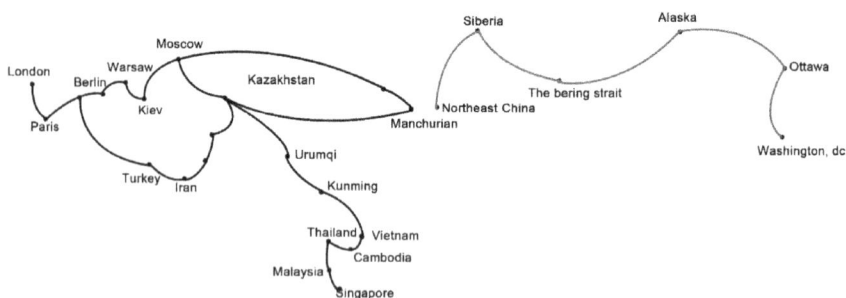

Figure 6.25   Global integration under SSR.

are 600 km apart and the trip is supposed to take only 30 min. One SSR train can leave every 30 s, carrying 28 people. The one-way fare is only $20. Therefore, low construction costs and moderate ticket prices make it possible to build global integration under the SSR system. This global integration process will start from the United States, pass through Canada to Alaska in the United States, cross the Bering Strait, enter northeast China from Siberia in Russia, cross major cities in China through major routes, then enter central Asia and the Middle East from Urumqi, and finally enter Europe from West Asia. The expected global integration plan of SSR is shown in Figure 6.25.

6.3 Summary

People's pursuit of speed is eternal. Nowadays, the speed of an HSR can reach more than 350 km/h, but it is far from enough for people who state that "time is money." Therefore, Elon Musk put forward the concept of "SSR" in 2013, subverting the traditional rail transit travel mode, more in line with today's fast-paced modern life. It has only been more than five years since the SSR system was put forward, but it has moved to practical testing from the concept design, which is closer and closer to the operation of the SSR system. With the maturity of SSR technology, more and more countries will participate in the research and development of SSR systems, making the commercial operation of the SSR system a reality.

Because the speed of the SSR system is far greater than that of the HSR, the development of the SSR system will have a significant impact on national and regional development strategies. The establishment of the

SSR system can realize the integration between countries and regions, eliminate the impact of geographical constraints between countries, and accelerate the exchange and cooperation of economy and resources among regions. Considering the overall situation of international development, the SSR system will have a far-reaching strategic impact because, on the one hand, strategically, it can ensure national security. On the other hand, due to low construction cost and moderate ticket price, there will be greater passenger demand to meet people's travel requirements. Therefore, considering the overall situation of global development, the development of the SSR system will have a very far-reaching impact on world politics and the economy. The SSR system can not only promote world integration but also realize a "global village."

Chapter 7

Case Studies of Super-Speed Rail (SSR)

After Elon Musk proposed the concept of SSR, it attracted the attention of experts and scholars. As early as the 1960s, some scholars began to study evacuated tube transport, and many scholars have conducted relevant tests and experimental research. Due to the frontier and strategic nature of the SSR system, many countries have invested a large amount of money doing further research. Currently, in addition to Musk's company, two other US companies, Virgin Hyperloop and Hyperloop TT, are also working on this goal. However, whichever company is the first to build the SSR with the highest commuting rate may not go as far as Musk expects. While Elon Musk suggested ultra-high iron capsules suspended in high-pressure gas, both companies tend to use capsules that are magnetically levitated, which works more like maglev. Therefore, this chapter makes a comprehensive analysis of SSR systems in the United States, China, France, Russia, Japan, India, Saudi Arabia, and Canada, so that readers can further understand the current research status of SSR systems.

7.1 Case Study of American SSR

The concept of SSR was proposed by American Elon Musk, so the United States became the first country to study SSR. In 2013, Musk came up with the idea of SSR as a rapid transit system. The new high-speed transportation system would put aluminum capsules into hundreds of kilometers of low-pressure tubes and launch them like artillery shells to their

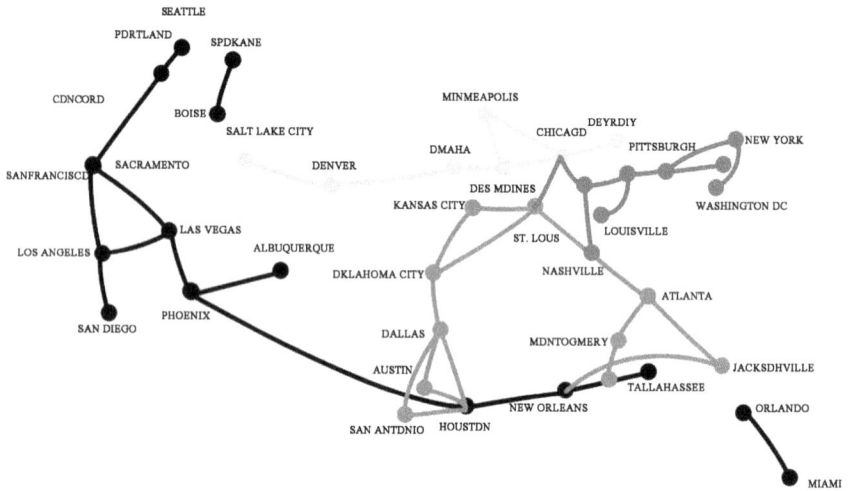

Figure 7.1   Musk's "Hyperloop."

destinations, which would transport passengers back and forth. Musk first came up with the SSR, which he believes will be able to transport passengers over long distances at speeds of up to 1,200 km/h. Currently, Musk is building a 400 km SSR line between San Francisco and Los Angeles that can transport passengers in 35 min and is expected to carry passengers by 2030. On August 12, 2013, Musk proposed another ultra-fast intercity transportation project, "Hyperloop," as shown in Figure 7.1.

"Hyperloop" is an ultra-fast urban transportation system powered by solar energy. At present, it takes US passengers an hour to travel from Los Angeles to San Francisco by air, but the appearance of the SSR could change the concept of traditional transportation. On the ground, passengers can possibly travel from Los Angeles to San Francisco by SSR in 30 min. Due to the speed of SSR, several American companies began to invest in and study the SSR project. At present, the main ones are Boring Company, Virgin Hyperloop (which ended its SSR passenger research in February 2022 to develop SSR freight research), Hyperloop Transportation Technologies, and Arrivo (which ended SSR research in November 2017 and shifted to maglev high-speed train research).

Faster than civil aviation, when such a means of transportation becomes a reality, it will undoubtedly be an important breakthrough in the history of human transportation. It is exciting to think about it. The SSR concept, proposed by Musk in 2013, uses vacuum tubes and magnetic levitation technology to enable transportation at thousands of kilometers

Figure 7.2 Exterior of Virgin Hyperloop One cargo Bay.

Figure 7.3 Interior of Virgin Hyperloop One cargo Bay.

per hour and was once considered the future of transportation. At present, SSR is gradually moving from theory to concept, and from concept to reality. For example, on July 25, 2020, to promote further development of SSR, the United States Department of Transportation and the Committee on Non-Traditional and Emerging Transportation Technologies (NETT) released a clear guidance document on the regulatory framework of SSR, which clearly states that "SSR provides a unique opportunity to lead the world in 21st-century transportation." Figure 7.2 shows the exterior of the Virgin Hyperloop One cargo bay and Figure 7.3 shows the interior of the Virgin Hyperloop One cargo compartment.

7.1.1 *Research background of American SSR*

The United States is the second largest country in the Americas. Its territory includes mainland United States, Alaska in the northwest of North America, and the Hawaiian Islands in the central Pacific Ocean. The total population of the United States is about 333 million. Although the Central Great Plain is a land suitable for human production and life, with flat terrain, fertile land, suitable climate, and sufficient water, the population of this land is very small, only about 5.5 million people, while the coastal area is vast enough to bear the current population of the United States. Therefore, the population and major commercial activities of the United States are mainly distributed in the coastal areas. The vast land and the uneven distribution of population result in the transportation mode of Americans being mainly by road and air. Short-distance commuting depends on road transportation, while long-distance travel depends on air transportation. At present, the United States has a nationwide highway system, and every city has relatively complete airport facilities. In contrast, the development of railways is stagnant. Rail transportation is not only expensive up front but also slow, so many rail companies have gone out of business in the face of the impact of aviation. However, with the emergence of SSR, a mode of transportation that replaces cars and planes appears in people's vision. It makes up for the shortcomings of existing modes of transportation for medium-distance travel and has achieved breakthrough development in terms of safety, comfort, environmental protection, and economy. Therefore, taking the traffic behavior between San Francisco and Los Angeles in California as an example, this paper expounds on the research background and significance of SSR in the United States.

The corridor between San Francisco and Los Angeles in California is one of the busiest transportation lines in the west of the United States, with a total distance of about 615 km. At present, the two main transportation modes are highways and air. If the SSR is built, there will be three modes of transportation: Road, Air, and SSR:

Mode 1: Road travel. If you take a road trip, it typically takes 5 h and 30 min, and costs about $115.

Mode 2: Air travel. Air travel takes an hour and 15 min, and costs about $158.

Mode 3: SSR travel. A trip on the SSR takes 35 min, and the SSR leaves every 30 s. Each trip carries 28 people and costs about $20. With

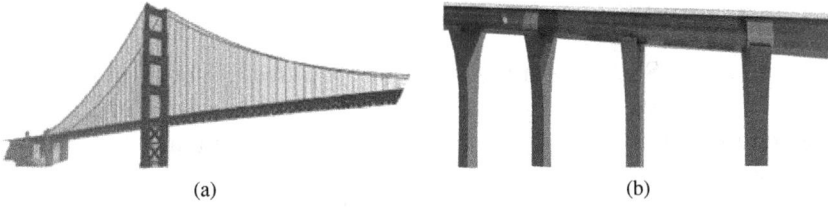

(a)   (b)

Figure 7.4   Design of the SSR between Los Angeles and San Francisco. (a) The SSR's bridge over the sea. (b) The SSR's operating pipeline.

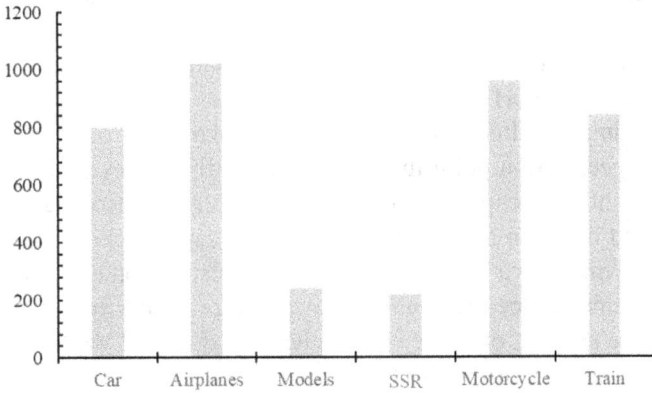

Figure 7.5   Energy consumption cost of transportation modes between LA and SF.

a planned construction cost of $6 billion between Los Angeles and San Francisco, a one-way ticket could be priced at $20 each. The design of the SSR between Los Angeles and San Francisco is shown in Figure 7.4.

Therefore, under various constraints, only SSR can improve travel efficiency and time. Especially with the requirements of energy conservation and environmental protection, as well as adverse factors such as global energy shortage, energy consumption is also an important factor to consider for emerging transportation modes. In terms of energy consumption, SSR is far lower than all other transportation modes. The energy consumption cost of various transportation modes between Los Angeles and San Francisco is shown in Figure 7.5. As can be seen, the estimated energy consumption cost of SSR from Los Angeles to San Francisco is lower than that of any current transportation mode and far lower than that of some traditional transportation modes.

7.1.2 Research results of American SSR

Hyperloop (or SSR) is a high-speed intercity transportation system that places passengers and cargo inside transport pods operating in a decompressed pipeline system. The transportation speed can reach 1,300 km/h, and the annual transportation capacity will reach 15 million passengers. To reduce friction between the pods and the rails and air, the Hyperloop may use magnetic levitation technology to levitate the pods in the depressurized tubes, using very little energy to propel the pods at high speeds. The SSR dream has sparked a lot of entrepreneurial enthusiasm. Two of the biggest players are Hyperloop Transportation Technologies (HTT) and Hyperloop Technology (now renamed Hyperloop One). Three months after Hyperloop Alpha was announced, German-born entrepreneur Dirk Ahlborn established HTT.

According to Musk's original vision, solar cells on a pipe would power an acceleration point that would be configured every 40 or 50 mi. At those points, the capsule would gain additional electromagnetic acceleration. And between acceleration points, the capsule can slide inertially without slowing down too much because the suspension and low air pressure in the pipe reduce friction. Further, relying on solar energy to provide an explosive amount of power to accelerators does not work everywhere and in all weather conditions. The phases of development of the SSR are given in the following:

Phase 1: The idea of the SSR system. To reduce friction between the pods and the rails and air, the SSR might use magnetic levitation technology to levitate the pods in depressurized tubes, using very little energy to propel them at high speeds. The SSR is designed to use magnetic levitation and a low-pressure track to achieve unprecedented high speeds. The cars will move on a closed track where the air is nearly drained and there is virtually no air resistance, much like a jet plane flying at extremely high altitudes, according to HTT. The remaining air in the track will be pumped to the back of the car through an air compressor to propel it forward. This allows the train to reach speeds of 1,220 km/h with minimal energy consumption.

The SSR concept was born when Musk first proposed it in August 2013. On his SpaceX website, he published a 58 page white paper called *Hyperloop Alpha*, which listed the Hyperloop as the fifth mode of human transportation (the others being planes, trains, automobiles, and ships).

According to Hyperloop One, passengers on the SSR will not experience acceleration because the acceleration is gradual, and when it reaches a top speed of 1,230 km/h, the journey between San Francisco and Los Angeles will take just 30 min. The journey would take six hours by car and a full day on a conventional train.

Basic parameters of SSR: The speed of the SSR is 3 times greater than the fastest bullet wheel rail high-speed train and 2 times greater than the speed of an airplane. The basic parameters of SSR in the United States are as follows:

Parameter 1. Running speed: above 1,200 km/h.
Parameter 2. Running line: elevated + straight line.
Parameter 3. Operating principle: vacuum pipeline + magnetic levitation.

Phase 2: Unmanned test of SSR system. The SSR system in the United States has been operated in several passenger-free tests, among which there are three representative tests as follows:

The first test — Acceleration test: On May 11, 2016, Hyperloop One was successfully tested in the Nevada desert north of Las Vegas, USA. The 3-m-long experimental "tackle" accelerated to 640 km/h in 2 s and slowed to a stop when it hit a sand pile 91 m away. Brogan BamBrogan, the co-founder of Hyperloop One, revealed that this was mainly conducted for testing the hardware and system, and now the goal is to accelerate to 644 km/h in 2 s. Hyperloop One envisions the SSR traveling through a closed vacuum or low-pressure tube, reaching speeds of up to 1,230 km/h. An SSR test vehicle is shown in Figure 7.6.

Figure 7.6 SSR test vehicle.

This test of Hyperloop One represents an important step forward for the project and proves that SSR is not a pipe dream. In particular, the success of the first outdoor test of SSR's propulsion system has given the world a glimpse of the "future of transportation on Earth."

The second test — Vacuum test: In May 2017, Hyperloop One completed its first "full vacuum" test of the SSR system at its beta track in North Las Vegas, USA. For the test, which took place in the desert, the team built a 4.83 km track with a cable pylon that would levitate the supporting pipe above the ground. The test, which focused on "full vacuum conditions," was also relatively short, running for just 5.3 s in the vacuum tube and reaching a top speed of 113 km/h. Using the full-size passenger cabin, XP-1, which is 8.7 m long and 2.4 m wide, the test car ran almost the entire distance, including a 300 m acceleration process, reaching a top speed of 310 km/h. A schematic diagram of SSR's first "full vacuum condition" test process is shown in Figure 7.7.

The third test — Speed test: Operating in a full vacuum, SSR's operating environment is equivalent to an airplane flying at an altitude of about 60,000 m. At such altitudes, there is little air resistance, ensuring faster flight speeds. Hyperloop One increased the speed of the SSR to 402 km/h in this test (Figure 7.8). This time, the test car took 300 m to accelerate on a 500-m track. In this experiment, all aspects of the SSR system, from the engine to electronics to the vacuum pump and the magnetic suspension mechanism, in the process of testing were running well. The future goal is a speed of more than 402 km/h in the controllable test environment, but

Figure 7.7   Schematic diagram of SSR's first "full vacuum condition" test.

Figure 7.8 Operation process of SSR test vehicle.

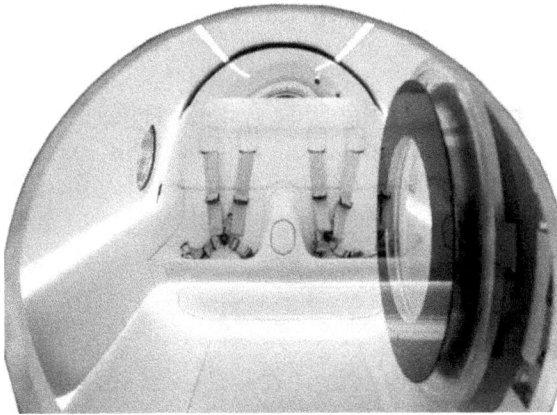

Figure 7.9 SSR passenger cabin.

it needs a longer test track, the maximum speed can reach 805 km/h in the future.

Phase 3: Passenger-carrying trials of the SSR system. On October 7, 2020, Virgin Hyperloop announced that their superfast pipeline train had completed its first passenger-carrying test. In the Las Vegas desert, Virgin Hyperloop's SSR had its first test with people on board, with Josh Giegel, Virgin Hyperloop's Chief Technology Officer, and Sara Luchian, Director of the passenger experience. The SSR passenger cabin is shown in Figure 7.9.

The capsule for this SSR passenger test can hold two passengers and is equipped with the five-point harness commonly used in racing cars. Virgin Hyperloop hopes the final capsule will hold 23 passengers. The test was carried out in a 3.3-m-diameter, 500-m-long tube, and it took 15 s for the capsule to reach the finish line, which translates to a speed of 172 km/h. That is not quite as fast as a high-speed train, and far from Virgin Hyperloop's envisioned speeds of more than 1,000 km/h, reportedly due to concerns for passenger safety and comfort, but also because the 500 m tunnel is not enough for full acceleration. But Virgin Hyperloop holds this test as an important milestone for SSR, as it transported people in a vacuum, something no other agency has yet done. Virgin Hyperloop has now conducted more than 400 unmanned tests in the pipeline. The highest speed recorded to date is 386 km/h, reached in a 2017 test. For faster-manned tests, a longer vacuum pipe would have to be built. For Virgin Hyperloop to accelerate to the envisioned 1,126 km/h, it would need at least another 2 km of orbit.

Phase 4: Construction of the SSR system. Hyperloop Transportation Technologies (HTT) began construction of the "SSR" passenger compartment (Figure 7.10) on March 21, 2017 in preparation for SSR operational

Figure 7.10   The passenger compartment of the "SSR" built by HTT.

testing. The SSR train is scheduled to carry 164,000 passengers a day and will depart every 40 s. The new transport system will be able to accelerate to 1,220 km/h, faster than the top speed of most aircraft. On May 12, 2017, Hyperloop One conducted the first full-scale test of its SSR technology in a vacuum, using magnetic levitation technology to reach speeds of 113 km/h at a test site in Nevada. It reached 310 km/h in a July 2017 test.

After years of research and design, Florida mechanical engineer Daryl Oster received a patent for his invention of the evacuated tube transportation (ET3) system in 1999. ET3's first choice for the body is maglev. The patent describes a 100-kg MoPodTM vehicle that can be driven directly from one's home to the SSR station and then into a tube to carry the super train at high speeds. The MoPodTM vehicle has a diameter of 1.3 m and a length of 4.8 m. It can carry 6 people or 300 kg at full load, which is the best ratio of efficiency and cost. In addition, there is a kind of MoPodTM vehicle that is more suitable for family use, which can usually carry two adults and part of their luggage. The MoPodTM vehicle consumes only 5 kW of energy, is powered by an internal combustion engine, and is fueled mainly by renewable energy. Parameters of the MoPodTM vehicle are shown in Table 7.1.

On October 19, 2018, Virgin Hyperloop One said it had identified a "viable route" in the United States along the Interstate 70 highway that runs through Missouri, connecting Kansas City, Columbia, and St. Louis (Figure 7.11). Missouri is an advantageous location for building the SSR. Topographically, the U.S. state of Missouri is largely flat and hilly, which would greatly reduce construction costs. If the vacuum line is not straight, it can lead to a "roller coaster" effect.

Virgin Hyperloop One claims that the first passenger-carrying SSR will be built in Missouri by the mid-2020s (Figure 7.12). Black & Veatch, an engineering firm based in Kansas City, has conducted a feasibility

Table 7.1   Parameters of the MoPodTM vehicle.

No.	Indicators	Parameter
1	Weight	100 kg
2	Diameter	1.3 m
3	Long	4.8 m
4	Load	6 people
5	Full weight	300 kg

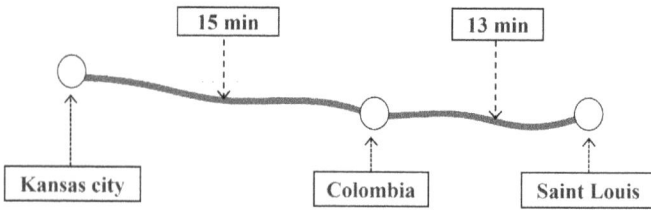

Figure 7.11   SSR route in Missouri, USA.

Figure 7.12   SSR test pipeline in Missouri, US.

study on the project and concluded that it would be practical to build the SSR in Missouri, with linear infrastructure costing 40% lower than global HSR, with a minimum cost of around $7 billion to $10 billion. The study also said the cost per drive of the SSR system would be less than the cost of gas for the same travel distance. Driving between Kansas City and St. Louis takes three and a half hours while traveling by SSR is expected to take just 28 min. And traveling from Kansas City or St. Louis to Columbia by car takes nearly two hours while traveling by SSR is expected to take only 15 min. The HTT experimental site is shown in Figure 7.13.

7.2 Case Study of Chinese SSR

By the end of 2021, the world will operate 50,000 km of high-speed rail, including a Chinese wheel-rail system of more than 40,000 km, thus ushering a "high-speed rail era." By 2030, China will realize basic connectivity both inside and outside the country, open inter-regional high-speed rail links between provincial capitals, and have fast access to prefectures and cities and basic coverage of county areas. China is standing at the

Figure 7.13   HTT experimental site.

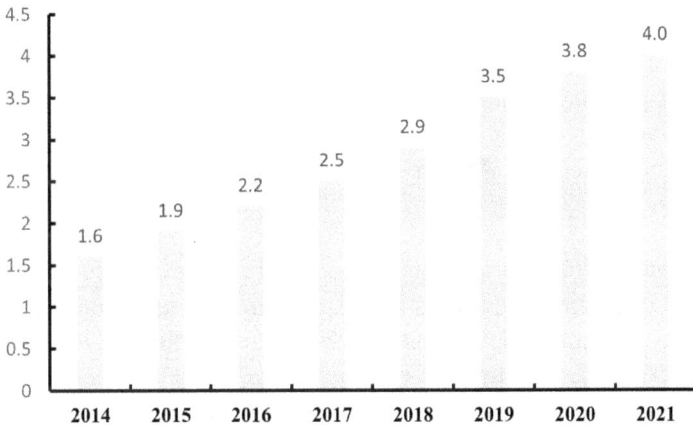

Figure 7.14   China's unified high-speed rail mileage (in units: 10,000 km).

forefront of the world's railways, working with other countries to meet global challenges, and leading the era of high-speed rail. In the arena of super-speed rail, China has also achieved good results.

7.2.1 *Research background of China's SSR*

China is located in the northern half of the Eastern Hemisphere, eastern Eurasia, and the west coast of the Pacific Ocean. China's high-speed rail network has grown from 9,356 km in 2012 to more than 40,000 km in 2021, as shown in Figure 7.14. The total length of high-speed in 2021 has more than quadrupled, putting it firmly in the first place in the world. China's "four vertical and four horizontal" high-speed rail network was completed ahead of schedule and "eight vertical and eight horizontal" high-speed rail network uses an encryption formation. In the era of rapid development of railway construction, China's technological break-throughs in high-speed rail are amazing. As a new concept in the post-high-speed rail era, super-high-speed rail represents the most advanced and cutting-edge technology in the world, and China is researching the same with much energy.

The idea for a "vacuum pipeline transportation system" was first proposed by mechanical engineer Daryl AOSTE in the 1990s. Elon Musk, CEO of Tesla, the US electric car company, fleshed out the concept with the idea of "super-high-speed rail" and contributed more design details to this transportation concept. At present, many countries in the world are stepping up the development of maglev trains. For example, Japan is planning a commercial operation line for cryogenic superconducting maglev trains, and the United States is developing super-speed trains with a speed of 1000 km/h. After entering the 21st century, Chinese scholars began to study "vacuum pipeline transportation system" one after another. The following are the phases of development of the SSR in China.

Phase One — The idea of a super-speed rail system: In 2004, Southwest Jiaotong University proposed a high-speed, high-temperature, and super-conducting magnetic levitation (HTS) Transportation System with a capacity of 600 km/h and above, thus beginning the exploration of HTS Maglev Vehicle Engineering. In 2011, Professor Zhao Yong of Southwest Jiaotong University led a team to develop the world's first "vacuum tube maglev experimental system." This is the world's first combination of vacuum tube, magnetic suspension, and linear drive comprising the complete vacuum tube test equipment. In 2013, under Zhao's leadership, his team developed the first high-temperature superconducting magnetic levitation (HTS) Test Loop, and later added a vacuum pipeline to become the world's first vacuum pipeline transportation (ETT) system.

Phase Two — Testing of the super-speed rail system: In 2014, Southwest Jiaotong University completed a prototype test for the ultra-high-speed maglev train, but the experimental loop has a radius of only 6 m and the test vehicle's top speed was only 50 km/h. In 2015, the second-generation high-speed ring was completed. The team laid tracks on the wall of the pipe to form a "wall-mounted" maglev train, which can effectively solve the centrifugal force problem caused by the small track radius in the laboratory. The average speed of the experimental vehicle was raised to 82.5 km/h under normal pressure. On September 3, 2017, the National Super-High-Speed Rail Laboratory of Beijing Jiaotong University constructed a 1 km super-high-speed rail test line at a pilot plant in Yantai, Shandong Province, to strengthen research on high-speed maglev transportation technology with vacuum pipelines. China will strive to achieve major breakthroughs in super-high-speed rail technology and actively promote the process of industrialization.

Phase Three — The vision of a super-speed rail system: The China Aerospace Science and Industry Corporation will develop a high-speed flying train in two steps. On August 29, 2017, the China Aerospace Science and Technology Corporation announced in Wuhan that it had started research and development on a 1,000 km/h "high-speed flying train" and would develop super trains with maximum operating speeds of 2,000 km and 4,000 km. On August 30, 2019, the China Aerospace Science and Industry Corporation launched a study on "high-speed flying trains," a new generation of vehicle, which combines supersonic flight technology and rail transit technology, devoted to achieving supersonic "near-earth flight" by using superconducting magnetic levitation technology and a vacuum pipeline.

7.2.2 Research results of China's SSR

At present, China's high-speed rail mileage has exceeded 40,000 kilometers. The Fuxing EMU runs at a speed of 350 kilometers per hour, and the technology is leading in the world. Higher speed has been the unceasing pursuit of scientists all over the world. In order to lead the world, China's different teams for the super-high-speed rail research have achieved impressive accomplishments:

(1) High-temperature superconducting high-speed maglev project of Southwest Jiaotong University
Southwest Jiaotong University began to develop maglev technology in the 1980s. In 1997, it was approved by the National 863 Program project

Figure 7.15   High-temperature superconducting maglev vehicle "Century."

"High-Temperature Superconducting Maglev Vehicle," and the research on high-temperature superconducting maglev vehicles was carried out. In early 2001, the school developed the world's first manned high-temperature superconducting maglev experimental vehicle "Century" in Beijing as shown in Figure 1.5, attracting widespread attention. The high-temperature superconducting maglev experimental vehicle "Century" had a maximum suspension weight of 700 kg. It could only carry two people, but caused a sensation in the country.

In 2011, a team led by Professor Zhao Yong, Director of the Research and Development Center for Superconductivity and New Energy at Southwest Jiaotong University, developed the world's first high-temperature superconducting maglev train test system with a vacuum pipeline, which had a track diameter of 3 m and pipeline minimum pressure 2,000 Pa, using linear motor drive. In January 2016, Zhao Yong's team completed the construction of a second-generation high-temperature superconducting maglev train system with a vacuum pipeline, which operated in a "wall-mounted" mode for the first time, which means the 6.5-m-diameter circular track is laid on the circular metal wall to enable the maglev to travel along the wall at high speed (Figure 7.16). In May 2016, the system completed the first phase of commissioning, reaching a speed of 108 km/h. In October 2016, the system completed the second phase of commissioning, reaching a speed of 150 km/h. The second-generation system effectively increased the centripetal force of the maglev car running along the loop line, and prevented the maglev car from derailing along the

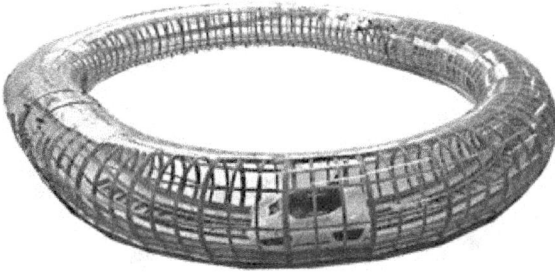

Figure 7.16 The second-generation vacuum tube HTS maglev train system.

tangential track, thus making the maglev car obtain higher running speed and safety stability.

In 2020, Southwest Jiaotong University, in conjunction with CRRC, China Railway, and others, carried out research on high-temperature superconducting maglev transport engineering prototype cars and test lines, constructed the integrated technology system of the HTS high-speed maglev, and carried out the suspension test, which indicates that HTS maglev technology has taken a key step from principle verification to engineering application. In January 2021, a silver-and-black bullet that resembled the front of a high-speed train but had no wheels appeared on the Chengdu Test Line, so Southwest Jiaotong University's "High-Temperature Superconducting High-Speed Maglev Project" (Figure 7.16) is gradually coming to fruition, with a design speed of 620 km/h and a full length of 165 m.

The "high temperature" of a high-temperature superconductor is the opposite of a low-temperature superconductor, which operates at −196°C. A superconductor is installed at the bottom of the prototype, and the track is a permanent magnet. When the liquid nitrogen lowers the temperature to −196°C, the resistance of the superconductor disappears, and the current creates a strong magnetic field in the superconductor, causing the car to naturally float. The maglev technology also has a "stick-in-the-mud" feature, allowing the system to automatically "pull back" the vehicle regardless of the direction of the force applied to it. Like nails in a plank, the train can only run along the track and never derail in operation. Figure 7.18 shows the HTS high-speed maglev prototype in the lab. Nitrogen is the main component of air, and the cost of obtaining liquid nitrogen is low. The "pinning" characteristic of this technology makes the suspension and

Figure 7.17   HTS high-speed maglev engineering prototype.

Figure 7.18   A high-temperature superconducting maglev prototype.

guidance system relatively simple without active control and vehicle-mounted power supply. The unique "pinning force" of the High-temperature superconductivity can not only ensure the safety of the vehicle but also maintain the stability of the vehicle body from side to side, which is difficult to achieve by any other means of transportation.

The lab's HTS high-speed maglev prototype showcases the self-levitation, self-steering, and self-stabilizing characteristics of HTS maglev transportation technology, which is a new type of rail transit mode with wide application prospects for future development. The high-temperature superconducting maglev technology is simple in structure, energy saving, with no chemical and noise pollution, safe and

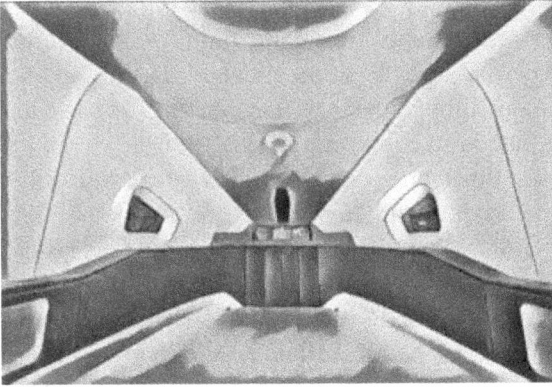

Figure 7.19   The console of the HTS high-speed maglev prototype.

Figure 7.20   Internal structure of HTS high-speed maglev prototype.

comfortable, with low operation cost, and especially suitable for high-speed and ultra-high-speed line operations. The console of the HTS high-speed maglev engineering prototype is shown in Figure 7.19. The internal structure of the HTS high-speed maglev engineering prototype is shown in Figure 7.20.

(2) "High-speed flying train" of China Aerospace Science and Technology Corporation
In 2017, the China Aerospace Science and Industry Corporation launched a study on "high-speed flying trains," which combine supersonic flight technology and rail transit technology. The new generation of vehicles

will use superconducting magnetic levitation technology and vacuum tubes to achieve supersonic "near-earth flight." China Aerospace Science and Technology has a wealth of practical experience and technical accumulation in major project systems engineering. It has the capability of model testing that large engineering simulations have, as well as international first-class supersonic vehicle design capability. All of these have provided an important foundation for the construction of the "High-Speed Flying Train" project. An operation diagram of China's super-high-speed rail is shown in Figure 7.21.

China's research plan for super-high-speed rail: The Chinese aerospace industry believes that the "High-Speed Flying Train" project will be gradually realized in accordance with the three-step strategy of increasing the operating speed to 1,000 km/h, 2,000 km/h, and 4,000 km/h. China's "high-speed flying train" concept is shown in Figure 7.22.

Step One, the subsonic stage — 1,000 km/h: Key technological breakthroughs will be achieved by 2020, and full-system demonstration and verification will be completed before 2023. A 600 km/h manned capacity will be created and a train with a maximum operating speed of 1,000 km/h will be developed. Then, we will construct a regional urban air train traffic network, forming a national metropolitan area with "Beijing–Shanghai–Guangzhou–Chongqing" as the core, as shown in Figure 7.23.

Figure 7.21 Operation diagram of China's super-high-speed rail.

Figure 7.22 China's super-high-speed rail.

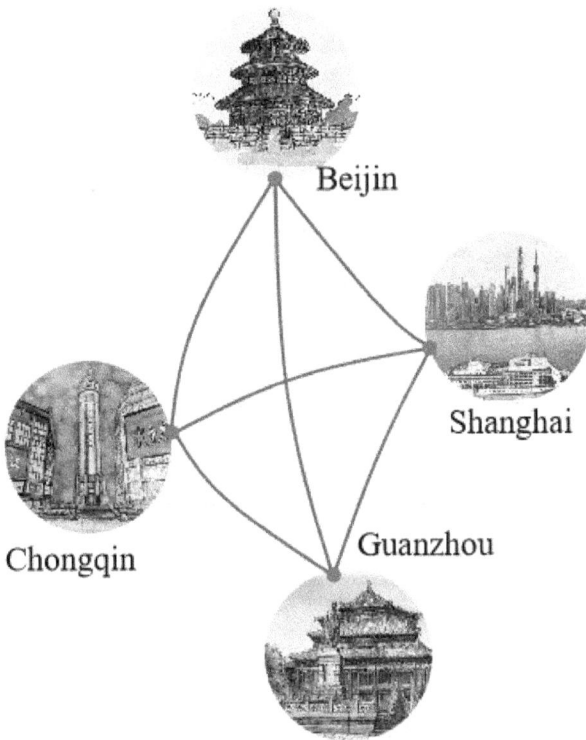

Figure 7.23 Regional urban air train traffic network.

Step Two, the hypersonic stage — 2,000 km/h: Before 2027, a train with a top speed of 2,000 km/h will be developed, and a flying train network will be built in urban agglomerations to form a national urban traffic network with provincial-level cities as the core, as shown in Figure 7.24.

Step Three, the supersonic stage — 4,000 km/h: We will develop a train with a top speed of 4,000 km/h, and construct and improve a national transportation network of "Belt and Road" flying trains, forming a world metropolitan area with capitals as the core, as shown in Figure 7.25.

Experimental test of China's super-high-speed rail: According to the Chinese aerospace industry, the speed of the high-speed flying train is 10 times greater than that of the traditional high-speed rail system, and five times greater than that of existing commercial airliners, with a maximum speed of 4,000 km/h, which is great progress in the pursuit of the fastest means of transportation. The high-speed flying train not only

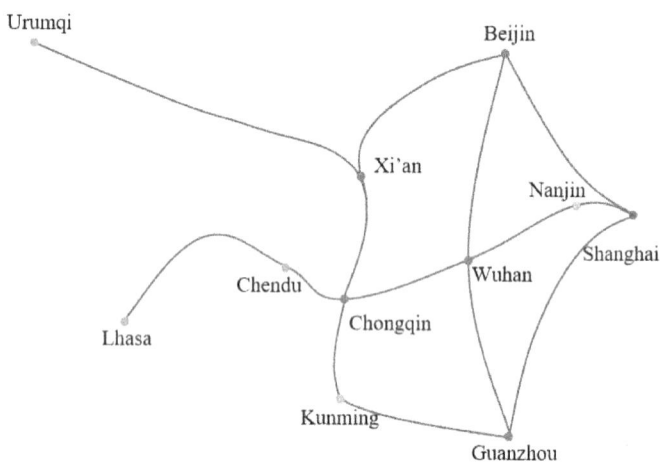

Figure 7.24 The flying train traffic network of city group.

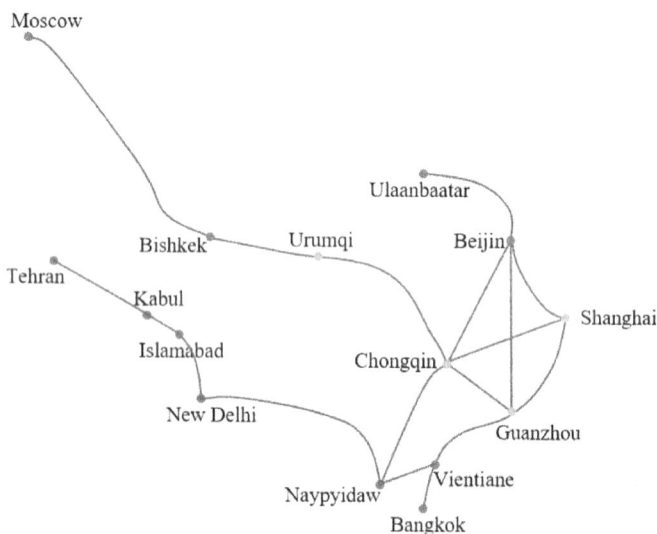

Figure 7.25 International air train traffic network.

shortens the space-time distance between cities but also has many advantages, such as not being affected by weather conditions, not consuming fossil energy, and connecting with urban subway seamlessly. The super-speed rail system is the development trend in the future

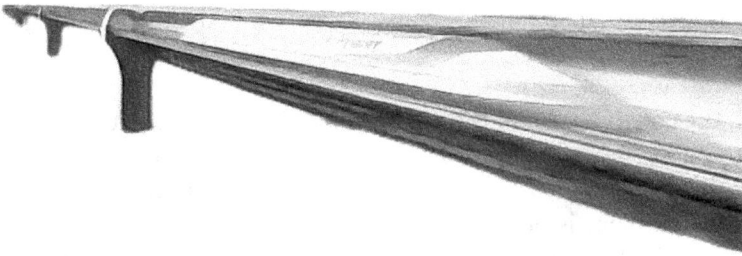

Figure 7.26 A sketch of China's super-high-speed rail system.

transportation field. A sketch of China's super-high-speed rail system is shown in Figure 7.26.

Research plan of China's super-high-speed rail: On November 4, 2021, the Chinese Academy of Aerospace Science and Technology conducted a test of ultra-high-speed magnetic levitation electromagnetic propulsion. During the test, the attitude of the vehicle was stable and the control of the converter system was accurate. The flight trajectory curve was in good agreement with the theoretical simulation, and the test speed reached 623 km/h. The test was successful, which proves the precision of the design of the suspension propulsion system and the coordination of each system, laying a solid technical foundation for the construction of a full-scale test line of the maglev transportation system for ultra-high speed and low-vacuum pipeline.

Problems of China's super-speed rail: China Aerospace Science and Technology (CAS) points out that a "high-speed flying train" accelerates slowly according to the speed that a person can bear, and it decelerates slowly according to the range that a person can bear. Because of its low acceleration, the passenger has no obvious sensation. How you feel depends not on how fast you are going, but on how fast you are accelerating. There is no such thing as harmful. The high-speed flying train has an acceleration process and a deceleration process. To ensure that it can safely stop and keep the human body in a comfortable state, the acceleration and deceleration of the distance, the middle of the cruise segment, and the site are planned and set carefully. At present, there have been some mature studies on the vacuum pipeline technology and magnetic levitation technology needed for the "super-high-speed rail" system, but there are still many difficulties. The technical difficulties mainly focus on three aspects: vacuum pipeline construction, car body traction technology, and maglev safety issues.

First, the construction of the vacuum pipeline: The problem is low-cost construction of the vacuum pipeline, that is, how one can realize and maintain a large volume of low-vacuum space at low cost. Another unsolved problem is how to build a platform, not only to facilitate passengers embarking and disembarking but also to maintain the vacuum state of the pipeline.

Second, the body traction technology: The "super-high-speed railway" needs to adopt the straight-line traction technology, but the efficiency of this technology cannot meet its power needs at present. According to the basic principle, as long as there is a positive force between the thrust and the resistance, the train will continue to accelerate, so there is no need for absolute vacuum, otherwise the project will be difficult. In terms of traction technology, China has a certain technical foundation through manned space engineering, but there are still challenges in manufacturing technology.

Third, the safety of magnetic levitation: Current magnetic levitation technology is not stable enough for the "super-high-speed rail." The impact of acceleration and deceleration on passengers' bodies and the change of acceleration when the train is turning is also one of the most important issues that researchers need to consider when developing and deploying this network of ultra-high-speed trains. After all, safety is always the most important issue.

   In short, while there are still technical problems with the technology, the quest for a faster, more comfortable, safer, more economical, and greener way to travel has not been thwarted. The progress from concept of super-high-speed rail to theoretical exploration and research, experimental proof of principle, demonstration operations, and to the comprehensive promotion needs the persistence of every generation. A few years from now, Maglev trains and super-high-speed trains may be as common and applicable as today's trains and high-speed trains!

7.3   Case Study of French SSR

France is a powerhouse of wheel high-speed rail (WHSR). On April 3, 2007, the French WHSR was on the test track, and the running speed reached 574.8 km/h, which not only created the myth of the WHSR but also made the WHSR the king of the ground. It was not until April 21, 2015 that the Japanese magnetic high-speed rail (MHSR) reached a speed of 603 km/h on the test track, and the position of the king of the ground was given to the MHSR. At present, the world has entered a "post-high-speed rail era," and super-speed rail (SSR) has become the focus of research and development in various countries. France does not want to be left behind in the field of high-speed rail and is also vigorously promoting the research and development of SSR.

7.3.1 Research background of the French SSR

France, or the French Republic, is located in Western Europe. It has a hexagonal shape with three sides facing the water. It is separated from Britain by the Lamanche Strait in the northwest. The plains of France account for 2/3 of the total area, and the main mountain ranges are the Alps, the Pyrenees, and the Jura Mountains. It is on the verge of four major seas: the North Sea, the English Channel, the Atlantic Ocean, and the Mediterranean Sea. From the perspective of geographical conditions, France's conditions are not conducive to the construction of a domestic SSR; but from the perspective of scientific and technological strength, France ranks fourth in scientific and technological development in the world. In 20 key scientific research areas, France ranks at the forefront of the world, and it is also innovative and adventurous in scientific research. France has advantages in aerospace, energy, material science, space technology, etc., and this strong scientific research strength is a powerful driving force required for the research and development of SSR.

HTT, the first company to start implementing Musk's vision, employs a large part of the workforce from NASA and Boeing. Another company called TransPod that is also competing in this area with HTT has many French employees. France's TransPod SSR system is under development, billed as the "fifth mode of transport." With plans to transport vehicles at speeds above 1,000 km/h, the system incorporates aerodynamics and propulsion systems. It reduces a lot of friction losses and transports passengers at a faster speed compared to trains, cars, and jets. The concept map of the French SSR is shown in Figure 7.27.

7.3.2 Research results of the French SSR

Facing the achievements of various countries in the development of SSR, France has also proposed its own SSR plan. France is expected to

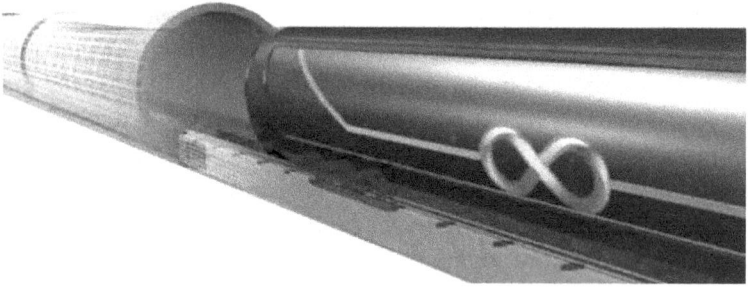

Figure 7.27   Concept map of the French SSR.

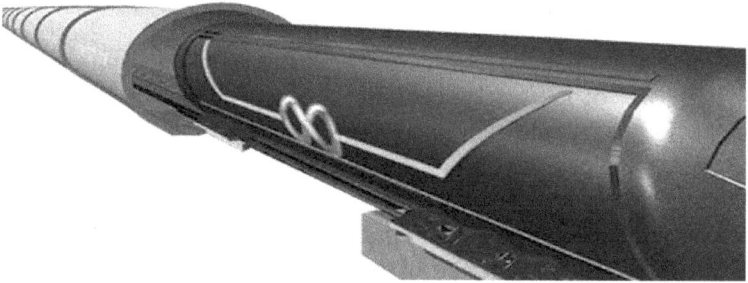

Figure 7.28   SSR system being tested in France.

complete the establishment of its own SSR system by 2025. The SSR system being tested in France is shown in Figure 7.28. The process is divided into three phases.

Phase 1: The idea of SSR system: The French hyperloop system uses air-cushion floating technology as the driving force. The air-cushion floating train first appeared in France. The technical pioneer in this field is the French scientist Jean Bertin. On March 5, 1974, the air-cushion train he developed created a record of an average speed of 417 km/h, and the instantaneous top speed was 430.4 km/h. Compared with the current HSR using magnetic floating technology, the French air-cushion floating technology can allow the train to form a pipeline air-cushion effect under the action of the air cushion, so that the train can float to the greatest extent. Under the action of the air cushion, the train can maintain a straight forward direction.

Figure 7.29   French SSR test track.

Phase 2: Testing of the SSR system: SSR represents more than just focused acceleration and record-breaking high speeds. What really needs to be done is the creation of an efficient and safe system with an unparalleled travel experience. In 2013, HTT Corporation set out to address the most pressing issues in transportation: efficiency, comfort, and speed. In 2018, this goal was basically achieved. Therefore, HTT opened the European SSR Research Center in Toulouse, France. This is because Toulouse is not only the fourth largest city in France but is also home to several transportation and aviation giants, including Airbus.

Phase 3: Construction of the SSR system: In April 2018, the construction of Europe's first SSR test track began in Toulouse, France. HTT has shipped the first batch of track tubes to a research center in southwestern France, which is the first test track for passengers and cargo. The construction of the high-speed rail network was officially launched. The test track for the French SSR system is shown in Figure 7.29. The details of the pipeline system are given in the following.

Phase 1: The French SSR test, which includes a closed 320 m pipeline system, was completed in 2018. Cargo pipelines of the first batch have an inner diameter of 4 m and are designed for both freight and passenger use, as shown in Figure 7.30.

Phase 2: The French SSR test consists of a test pipeline system up to 1 km long on a 5.8-m-high bridge, completed in 2019, as shown in Figure 7.31. In particular, the completion of the first full-size passenger

(a)

(b)

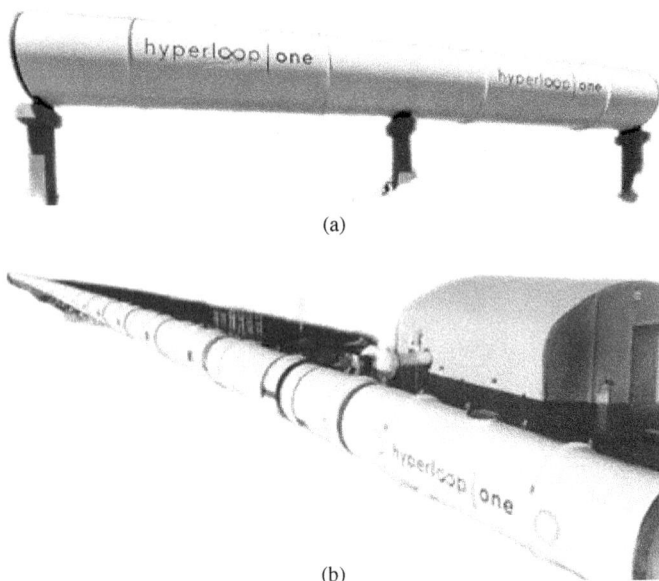

Figure 7.30   French SSR pipeline in transit. (a) Overhead pipeline. (b) Ground pipeline.

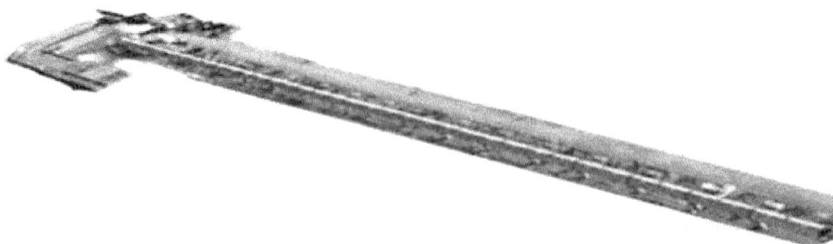

Figure 7.31   French SSR test route.

cabin in Toulouse, France, can be regarded as a milestone event, which means that SSR is no longer a concept, but has become an industry.

7.3.3 *TransPod for the French SSR*

On July 22, 2022, TransPod started building the world's leading ultra-high-speed ground transportation system (TransPod Line) and launched FluxJet, as shown in Figure 7.32. Based on breakthrough innovations in

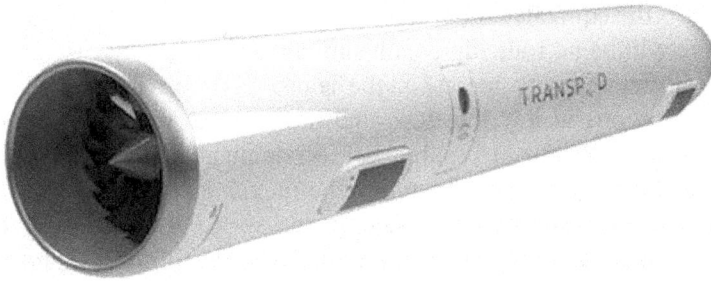

Figure 7.32 French SSR.

Figure 7.33 French SSR system.

propulsion and a clean energy system without fossil fuels, the FluxJet is an all-electric vehicle that is effectively a hybrid between an airplane and a train. FluxJet takes a technological leap in contactless power transfer and employs a new field of physics called long phase flux to travel at speeds in excess of 1,000 km/h in protected guideways, faster than jets and three times faster than high-speed trains.

(1) **TransPod system:** The SSR being developed in France will transport passengers and goods between cities at a speed of more than 1,000 km/h, and is a new generation of super-high-speed ground transportation that can be built around the world, as shown in Figure 7.33. SSR trains have a top speed faster than jet planes and travel in protected ground tubes that are not affected by the weather. The convenience of

SSR is the equivalent of a subway that departs every few minutes, and there will be no route diversion similar to airlines.

FluxJet will only operate on the TransPod Line, a networked system with stations in key locations and major cities with high-frequency departures designed to enable fast, economical, and safe travel.

(2) **TransPod's passenger system:** The maximum speed of Transpod's passenger transportation system is up to 1,000 km/h, and it can accommodate up to 54 passengers, 2 wheelchair spaces, and 4 luggage racks, with a payload of up to 10 tons, and the frequency of departure can be every 2 min. The French SSR Passenger System is shown in Figure 7.34.

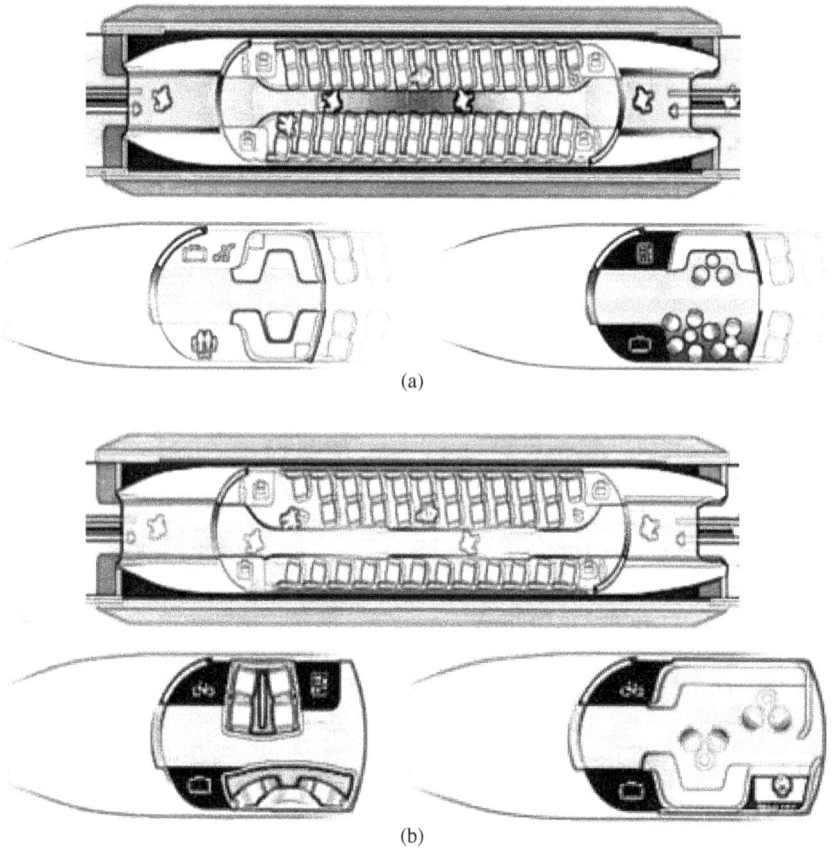

(a)

(b)

Figure 7.34 French SSR Passenger System. (a) Economy class. (b) Business class.

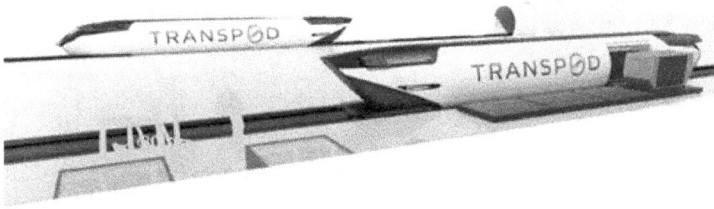

Figure 7.35   French SSR Cargo System.

Figure 7.36   French SSR test route in desert.

(3) **TransPod's cargo system:** On the freight side, TransPod's cargo terminal design allows fast and automated multimodal transport of containers and pallets that can be loaded on specialized pods, as shown in Figure 7.35. Traveling at speeds in excess of 1,000 km/h, the TransPod's all-electric system enables fast delivery with no carbon footprint. The cost per ton per km of TransPod is more affordable than air freight, allowing businesses to save on travel time and storage costs.

The innovative and cost-effective design of the French SSR is a new leap toward a clean, safe, economical, and comfortable mode of transport. And TransPod is working with the European Union and the U.S. Transportation Commission to develop this technology with high safety standards. The French SSR test route is shown in Figure 7.36.

7.4 Case Study of Russian SSR

Russia is the country with the largest land area in the world, and has a large demand for long-distance transportation. The SSR transportation

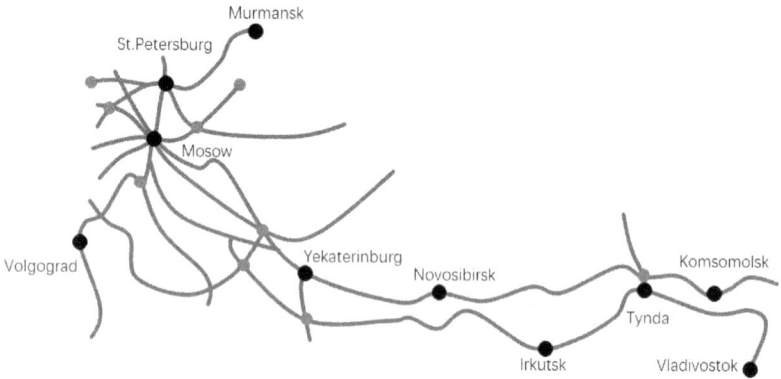

Figure 7.37   Distribution map of Russia's National Railway Network.

project will be a huge project. The distribution of the Russian railway network is dense in the south and sparser as it goes north. There is basically no railway distribution in the extreme north, and the overall trend is east–west with the territory. Since the European part has a large population and the economy is more developed than the Asian part, the European part of the railway network is denser than the Asian part. Figure 7.37 shows the layout of the Russian railway network.

7.4.1 *Research background of the Russian SSR*

From the perspective of Russia's existing railway network, the development of SSR throughout Russia and the world is a development trend, and it also has a great role in promoting Russia's energy transportation. The completion of the SSR will solve many development problems in vast areas of Russia, as shown in Figure 7.38. The project is proposed to connect with social programs, new fields, new energy resources, etc. The idea is based on the new technology of HSR transportation, which could promote Russia's tourism industry and make Russia a hub of world transportation.

7.4.2 *Research results of the Russian SSR*

In March 2015, Russia proposed an ambitious plan to build the SSR linking Moscow with Arras and London in the United Kingdom (as shown in

Figure 7.38 Eurasian–American integration.

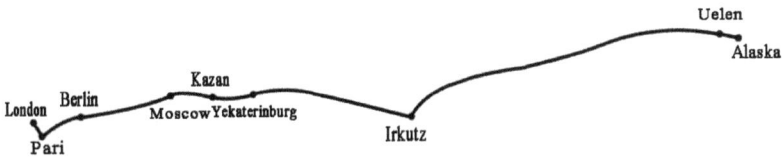

Figure 7.39 Planned route of the Russian SSR.

Figure 7.39), with a total length of 20,000 km. From London to Moscow to Alaska by train will no longer be a dream.

Russia's Trans-Eurasian Belt Development Program: The Russian Railways Joint Stock Company proposed the abovementioned plan called "Trans-Eurasian Belt Development." This line in Russia is similar to the Trans-Siberian Railway, the main line that traverses Russia from east to west, and will also pass through cities such as Yekaterinburg, Irkutsk, and Vladivostok. The railway will connect to the existing railway network in Europe, while also making it easier to travel to the Russian Far East. The rail network may extend to the Chukota region of the Russian Far East, across the Bering Strait to Alaska, and it will be possible to reach the United States by train from the United Kingdom, as shown in Figure 7.40.

Russia's SSR Program: In mid-May 2016, Russian Minister of Transport, Maxim Sokolov, said at a press conference in Sochi that Russia currently has technical reserves and is very interested in building large-scale transportation projects like SSR proposed by Musk. The first part of the SSR will connect Moscow and St. Petersburg, which are nearly 650 km apart, said Zaitsev, an official at St. Petersburg State Transport University. State-owned Russian Railways will consider a partnership

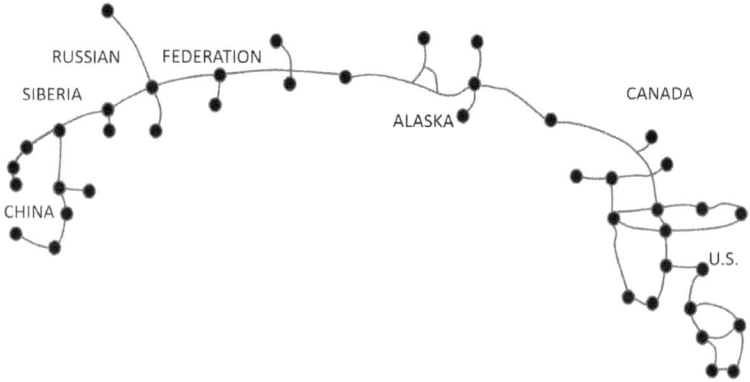

Figure 7.40   Eurasian–American railway map.

Figure 7.41   Russia prepares SSR test site.

with Hyperloop One, one of two U.S. companies developing the SSR project, from which Russia hopes to obtain technical support. Currently, they have formed a joint working group with Hyperloop One to study the possibility of jointly developing the SSR. The Russia SSR Test Site is shown in Figure 7.41.

Since 2016, the SSR has developed rapidly in Russia. The American SSR company Hyperloop One has signed an agreement with the Russian investment company Summa Group and the city of Moscow to study the construction of a transportation network in the city of Moscow, in the hope of improving the quality of life in the Moscow region. If the SSR is completed in the future, it is expected to transport containers from China to Europe within one day, as shown in Figure 7.42.

Figure 7.42   Russian SSR pipeline.

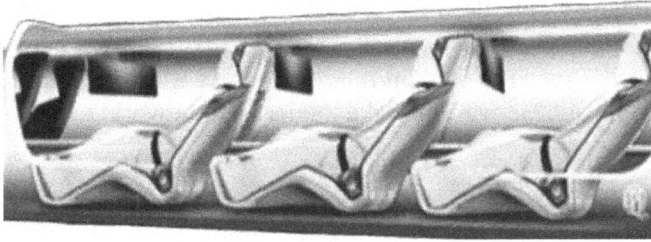

(a)

(b)

Figure 7.43   SSR Capsule System. (a) Russian capsule train interior. (b) Operation schematic diagram of the Russian capsule train.

In addition to the continuous advancement of pipeline laying and project engineering, Russia is also developing a "capsule system" supporting it. Hyperloop One signed an agreement with the Russian government to build a "capsule train" system for Moscow to support its existing transportation, as shown in Figure 7.43.

Sino-Russian cooperation SSR Program: At present, Russia is also actively cooperating with China in the field of HSR. China and Russia have reached an agreement to start an SSR project with a total investment of over 1 trillion yuan. China will help Russia build the SSR with a total length of 7,000 km. It is the Sino-Russian cross-border SSR. It is said that this SSR will cross China, Kazakhstan, and Russia. It will only take a few hours to travel by train from China to Russia after completion, which will greatly promote the economic and cultural exchanges between China and Russia.

7.5 Case Study of Japanese SSR

On October 1, 1964, Japan's wheel rail high-speed railway was put into operation. Japan was the first country to master the technology of the wheel-rail high-speed railway. On April 21, 2015, Japan's maglev bullet train reached a speed of 603 km/h, setting a new world record. For super-speed rail, Japan also wants to take the lead in the world.

7.5.1 *Research background of super-speed rail in Japan*

Since 1964, when the world's first high-speed railway opened in Japan, a new era of transportation has been ushered in. Japan is a mountainous island country and is located at the boundary between the Eurasian plate and the Pacific plate, near the Ring of Fire region in the Pacific, with volcanic and seismic activity being frequent. Although the objective conditions for the development of the super-speed rail are not very favorable, Japan's research on high-speed train technology is cutting-edge. However, in terms of geological conditions, it is still very difficult for Japan to build a super-speed rail with a speed of 1,200 km/h. It is not as simple as it is supposed to be. "Earthquake resistance" and other issues during the construction process need to be strictly considered. Figure 7.44 shows the wheel-rail high-speed railway or Shinkansen in operation in Japan.

7.5.2 *Super-speed rail research results in Japan*

Since Elon Musk proposed the idea of the super-speed rail, it has attracted attention from countries around the world, including Japan. In particular,

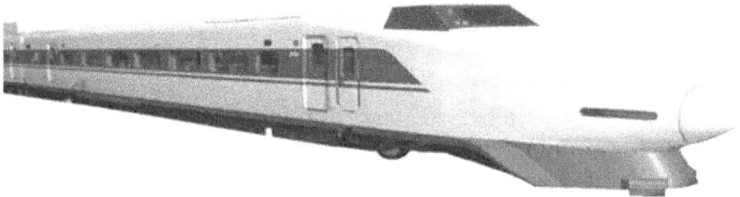

Figure 7.44   Japan's Shinkansen.

a team at Keio University in Japan is developing the basic technology for super-speed rail, and its influence is growing. Japanese researchers are actively developing super-speed rail technology and developing super trains. An SSR model developed by teachers and students in Japan is shown in Figure 7.45. Japan is also actively cooperating with foreign SSR companies in technology and is focusing on the development of a super-speed rail system with a speed of 1,200 km/h.

(1) **Japan develops a super-speed rail system:** Hitachi Railway, a subsidiary of Hitachi Corporation, has developed in cooperation with the U.S. super-speed rail transit technology company to integrate the physical ultra-signal operation and management infrastructure with the cloud-based railway signal system, and strive to achieve super-speed road tests, as shown in Figure 7.46.

(2) **Japan develops SSR track:** For super-speed rail driving technology, two types of research are being carried out: One is to use wheel

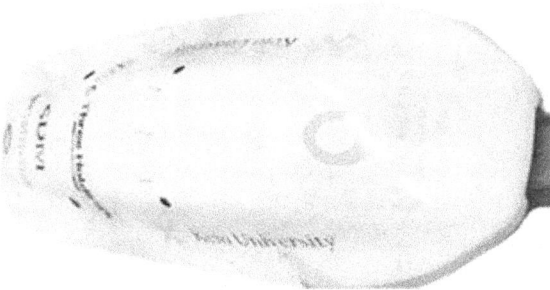

Figure 7.45 SSR model developed by teachers and students in Japan.

Figure 7.46 Concept map of Japanese super-speed rail.

driving like ordinary trains. If wheel driving is adopted, it needs to have the technology to reduce bearing load, friction, and noise even if it only exceeds the speed of the Shinkansen, as shown in Figure 7.47. The other is maglev.

(3) **Japan develops SSR trains:** The team at Keio University in Japan changed the magnetic field so that even non-magnetic materials like aluminum could have an electric current flowing through them, causing them to interact with magnets. By attaching a strong disc-shaped permanent magnet to the vehicle, a temporary electric current is passed through the track as it moves forward, creating a thrust force that levitates the vehicle, as shown in Figure 7.48.

Japan is currently experimenting with two types of propulsion to propel vehicles forward: magnetic propulsion and propeller propulsion.

Figure 7.47   Japan develops simulation system.

Figure 7.48   A magnetically propelled SSR.

Figure 7.49 SSR train with spiral propulsion.

The first is the magnetic force driving the vehicle. In 2017, the Keio University team attached permanent magnets to specific parts of the vehicle and rotated them, which interacted with the current flowing through the track to create the thrust of the super-speed rail vehicle.

The second was propeller propulsion. In 2019, a team from Keio University used propeller propulsion, which blows air backward like a propeller plane, to create the reaction force to move forward. Inspired by the fact that airplanes can fly at high altitudes where the air is thin, the Keio University team decided that the decompression of the pipe could also work as a propeller, as shown in Figure 7.49.

(4) **SSR construction in Japan:** Japan's JR Tokai plans to put the Central Shinkansen, a maglev train with a maximum speed of 550 km/h, into operation by 2027. The super-speed rail also requires a track and side wall, which can use a similar principle as the maglev central Shinkansen, as shown in Figure 7.50. But what to do with the decompressed pipe is a unique problem for the super-speed rail. If the pipes are decompressed, there will be a difference in air pressure between the inside of the vehicle and the pipes, and passengers cannot get on or off by simply opening the doors as in a normal train. For this purpose, a method of connecting the vehicle to the outside world at normal atmospheric pressure, as in the case of space station docking, needs to be introduced.

Figure 7.50   SSR construction in Japan.

7.6   Case Study of Indian SSR

India's roads and railways are well developed, ranking among the best in developing countries. Roads are mainly divided into three types, national roads, state-level roads, and border roads, connecting large, medium, and small cities all over the country to form a huge road network. Indian air routes now connect the world's five continents and are also among the leading in developing countries.

7.6.1   *Research background of SSR in India*

India is located in the northern hemisphere and belongs to the South Asian subcontinent. It is bordered by the Bay of Bengal in the southeast, the Arabian Sea in the southwest, the Indian Ocean in the south, and the Himalayan Mountains in the north. It is a maritime transportation hub of Asia, Europe, Africa, and Oceania. Since the 1990s, India's economy has made great progress, and the service industry has developed rapidly. The industry has formed a relatively complete system with strong self-sufficiency. India has become a major global exporter of services, such as software and finance. With the continuous improvement of living standards, the means of transportation are becoming more convenient. But for India, traffic is still a huge problem. India's underdeveloped transportation, coupled with a large population, means that trains are often packed densely with people, from the front to the rear, with almost no extra space. So, India wants to improve its transportation mode and solve the problem of people's mobility, which is very interesting for new modes of transportation, especially the super-speed rail.

7.6.2 Research results of super-speed rail in India

India has also tried many ways to solve the traffic problem, but to no avail. India claims to have built the fastest train in the country, Salute India, as shown in Figure 7.51, with a top speed of 180 km/h and the average speed of only around 160 km/h. At present, Indian officials say they have designated the ultra-fast, futuristic transport system as a public infrastructure project as the country seeks to build a super-speed rail line between New Delhi and Mumbai to improve the country's transport environment.

Super-speed rail, an ultra-fast futuristic transport system, has been added to India's public infrastructure projects. Super-speed rail is a kind of transportation tool with "vacuum steel tube transportation" as the theoretical core, which has the characteristics of ultra-high speed, high safety, low energy consumption, low noise, and low pollution. India is bidding

Figure 7.51   High-speed rail in India.

for a super-speed rail line to build a 1,300 km line from the Indian capital to Mumbai. As for the super-speed rail, the maximum speed will be 1,200 km/h. SSR construction in India is shown in Figure 7.52.

In 2017, the governments of the Indian states of Maharashtra and Karnataka began studying super-speed rail projects in the region. The super-speed rail project has been in development for a long time. Today, there are three planned super-speed rails in India, including the Bombay to Pune super-speed rail project. The Indian government and Branson's Virgin Group have invested $10 billion directly in SSR projects. A model of SSR is shown in Figure 7.53.

Therefore, considering the comprehensive advantages of the super-speed rail in all aspects, India is very determined to build a super-speed rail. The SSR line, India believes, can be completed between 2022 and 2025.

Figure 7.52   SSR construction in India.

Figure 7.53   The model of SSR.

7.7 Case Study of Saudi Arabian SSR

The kingdom of Saudi Arabia is a mystical, abundant resource, with a beautiful vast desert and unique social customs. However, few people know the high-speed railway is developing rapidly in this country. Saudi Arabia's railway total mileage is ranked 10th in the world and has a Mecca to Medina double-track electrified high-speed line, as shown in Figure 7.54, which is an astonishing achievement. It is also a rare high-speed railway in desert area in the world. As a country that developed later in the field of high-speed rail, Saudi Arabia has rapidly realized the new iteration of high-speed rail technology with the help of technology and financial support and occupies a place among the countries that have researched the super-speed rail.

7.7.1 Research background of Saudi SSR

Saudi Arabia is located in the Arabian Peninsula, bordering the Persian Gulf on the east and the Red Sea on the west. It borders Jordan, Iraq, Kuwait, the United Arab Emirates, Oman, Yemen, and other countries. The terrain of Saudi Arabia is high in the west and low in the east. Most of the territory is plateau. The west coast of the Red Sea is a long and narrow plain. The east is the Sierra Mountains. The terrain of the eastern mountains gradually declines until the east is flat and the desert is widespread. The north is the Great Nevada Desert and the south is the Rub al-Khali Desert. Although Saudi Arabia is willing to invest a large amount of money in super-speed rail, and there is a great demand for medium-distance traffic in the country, the extensive desert, hot climate, and complex topography bring great difficulties to the construction of Saudi Arabia's railway.

Figure 7.54 The Mecca–Medina high-speed railway.

7.7.2 *Research results of Saudi SSR*

Due to its unique political and economic environment, Saudi Arabia lags behind some other countries in the development of high-speed rail, but it has made great achievements in the field of super-speed rail through cooperation with foreign technology companies. In 2016, Saudi Arabia unveiled Vision 2030 for the country's long-term development and economic diversification. Renewable energy is a key development area in Vision 2030, and a super-speed rail that runs on solar power is a good fit for Saudi Arabia. Super-speed rail, in particular, is the catalyst that will enable all fourth-generation technologies to flourish in the Kingdom, while creating a vibrant society and thriving economy through desirable urban and high-tech clusters. Given in the following are details of the Saudi SSR system's development.

(1) **Saudi SSR system:** Through a partnership with Virgin Super-Speed Rail One, super-speed rail could become a reality in Saudi Arabia. The Vision 2030 super-speed rail capsule, as shown in Figure 7.55, is identical to the Virgin Super-Speed Rail One tested at its privately owned facility in Las Vegas. The pipe-based maglev system is theoretically capable of reaching speeds of up to 1,200 km/h, and the Virgin Super-Speed Rail One capsule has been tested at speeds of up to 386 km/h. Saudi Arabia is one of the few countries to have entered into a feasibility study with Virgin Super-Speed Rail One, whose ultimate goal is to work with organizations and countries to implement avant-garde transportation concepts and drive faster growth in the transportation industry.

Virgin SSR One extended its partnership with Saudi Arabia in 2019 by announcing plans to develop a new trial track in western

Figure 7.55   Vision 2030 SSR module.

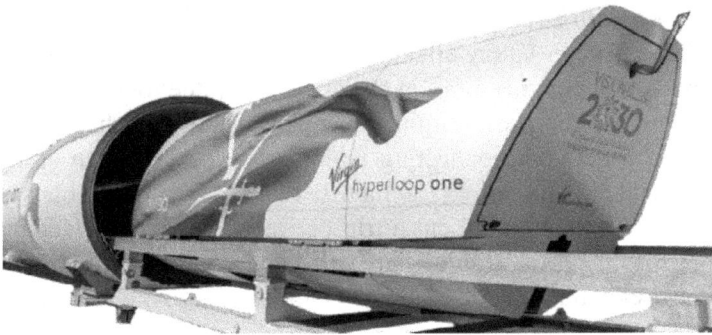

Figure 7.56 Saudi SSR system.

Figure 7.57 Through the tunnel.

Saudi Arabia, which will be the world's longest tunnel, to test the super-speed rail, as shown in Figure 7.56. Saudi Arabia, however, intends to expand the future of the super-speed rail transportation system and fully develop the super-speed rail, which is expected to transport people and goods in near-vacuum tubes at nearly the speed of sound.

(2) **Saudi SSR test:** Virgin Super-Speed Rail One has already built and operated a 500 m test tunnel in California since 2019, but the latest plan is to work with the Saudi government to build a 35 km tunnel to connect the King Abdullah Economic City north of Jeddah, greatly reducing the travel time between the two places as shown in Figure 7.57. In addition to the test tunnel, the facility will also be

equipped with a research and manufacturing unit, which it hopes will foster a "Silicon Valley effect" in the region and allow for the development of super-speed rail in Saudi Arabia in the longer term.

In the Middle East, Virgin Super-Speed Rail One had previously planned a network connecting major urban centers such as Kuwait City, Jeddah, Muscat, and Oman. If realized, Saudi Arabia's super-speed rail system could cut a trip between Riyadh and Jeddah from more than 10 h to 76 min. This will greatly improve the travel time and travel conditions of residents along the route.

7.8 Case Study of Canadian SSR

Canada is located in northern North America, and is bordered by the Atlantic Ocean to the east, the Pacific Ocean to the west, Alaska to the northwest, the mainland North America to the south, and the Arctic Ocean to the north. The climate is mostly sub-cold coniferous forest climate and humid continental climate, and the northern polar region has a polar long cold climate. This section will explore the possibility of super-high-speed rail development in Canada by understanding the background and current situation of Canadian railway development, as well as the government's demand for super-speed rail construction.

7.8.1 *Research background of SSR construction in Canada*

Canada's transportation is very developed, with water, land, and air transport being very convenient. Rail transportation is mainly freight, carrying most of the goods exported to the USA and overseas. A large proportion of the freight transported by rail each year is international trade in and out of seaports. But in terms of high-speed rail, Canada currently has no trains over 200 km/h and is the only G7 country not currently building any high-speed rail lines. Although trains exceeding 200 km/h were produced in Canada as early as 1967, the United Aircraft Corporation (now United Technologies) introduced a high-speed gas turbine rail passenger train. The appearance of the gas wheel train undoubtedly shocked people at that time greatly. In commercial operation in 1976, the roaring gas turbine train lived up to expectations, reaching a top speed of an astonishing 226 km/h, as shown in Figure 7.58. However, due to the early train operation, gas turbine trains began to have the issues of the brake system freezing in

Figure 7.58 High-speed gas turbine railway passenger train.

cold weather, suspension systems aging fast, and so on. As safety hazards continued to appear, gas turbine trains had to run at around 130 km/h most of the time. Eventually, the era of gas turbine trains came to an end in 1982 and the trains had to be officially retired, leaving Canada with no trains running at more than 200 km/h.

First proposed by Tesla CEO Elon Musk in 2013, the "super-speed rail" would be the fastest train in the world, reaching speeds of up to 1,000 km/h, rivaling an airplane. As a neighbor of the United States, Canada is also very optimistic about the prospect of the super-speed rail and will plan and build a corresponding super-speed rail project as well. In 2020, Canada planned to build a "super-speed rail" corridor from Calgary to Edmonton within five years. Toronto-based transportation technology company TransPod signed an agreement with the Canadian province of Alberta to develop and test the "super-speed rail" by building an infrastructure section of the Edmonton–Calgary transit system. Figure 7.59 shows the TransPod super-speed rail train concept diagram and Figure 7.60 shows the super-speed rail in the vacuum pipe envisaged by TransPod.

The super-speed rail system between Calgary and Edmonton by TransPod published the following timeline.

2020–2022: Preliminary preparatory work, field research, and so on;
2020–2024: Carry out engineering design and determine the construction process;
2024–2027: Construction and speed testing;
2030: Completed and operational.

Figure 7.59   TransPod's concept image of SSR train.

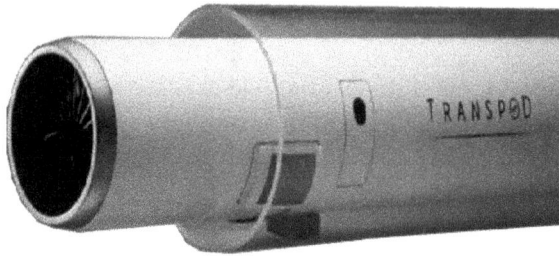

Figure 7.60   SSR in a vacuum tube.

7.8.2 Research results of SSR construction in Canada

The current major research results on super-speed rail in Canada are mainly provided by TransPod. In 2022, TransPod said it had begun preliminary construction of a pipeline that could travel from Calgary to Edmonton in 45 min. On July 22, 2022, TransPod Line, the start-up that is building the world's leading ultra-fast ground transportation system to disrupt and redefine passenger and freight transportation, launched FluxJet, an industry-defined innovation that is transforming the way we live, work, and travel.

(1) **SSR train in Canada:** Based on radical innovations in propulsion and fossil fuel-free clean energy systems, FluxJet is an all-electric vehicle that is essentially a cross between an airplane and a train, as shown in Figure 7.61. Flexjet has made a technological leap in contactless power transmission and employs a new physics field called

Figure 7.61   TransPod introduced FluxJet.

Figure 7.62   TransPod FluxJet in operation.

face flux to travel at over 1,000 km/h in a protected guide rail, faster than a jet and three times faster than a high-speed train.

(2) **SSR system in Canada:** FluxJet will operate only on the TransPod Line, a networked system with stations in key locations and major cities with high-frequency departure features designed to enable fast, economical, and safe travel, as shown in Figure 7.62. The ultra-fast surface transport system promises radical innovation in propulsion, compatibility with clean energy sources, and a technological leap in contactless power transmission, opening a new physics field called Veillance Flux.

(3) **Canadian SSR pipeline:** FluxJet uses magnetic levitation technology to eliminate rolling resistance. In this case, the train will reach a high speed when pass through the high-speed section. The "landing gear"

(a)

(b)

Figure 7.63   SSR train in Canada. (a) Operating vehicles. (b) Vacuum piping.

will not be repossessed until the pod reaches a speed of at least 300 km/h and is ready to accelerate to a cruising speed of around 1,000 km/h. Figure 7.63 shows a super-speed rail train in Canada.

(4) **SSR cabin in Canada:** The TransPod visually locates the Veillance Flux cross-section, scans the front pipe, and adjusts the pod's position in the pipe based on what is about to happen. Each 25-m-long pod, which can hold up to 54 passengers or carry a certain amount of time-sensitive cargo, will have four "levitating engines" extending from its upper and lower diagonal suspension arms. The pod will carry some onboard battery reserves, but when it is going very fast, TransPod will extend out a non-contact power pickup unit to receive power from the pipes.

As the super-speed rail continues to grow, it has also encountered some skepticism. Some super-speed rail critics have focused on the fact that the experience of passengers riding in a cramped, airtight, window-less capsule inside a sealed steel tunnel might be stressful and fraught with risk, mainly for the following reasons: The super-speed rail is subjected to tremendous acceleration forces, with high noise, vibration, and possible collisions due to air being compressed and transported around the capsule at nearly the speed of sound. Even if the tube is initially smooth, the ground may move with seismic activity. At high speeds, even small devia-tions from the straight path may add considerable chattering, making emergency handling tricky when faced with practical and logistical prob-lems of safety due to equipment failures, accidents, and emergency evacuation.

7.9 Summary

High speed is the eternal pursuit of mankind. However, the SSR system represents not only high speed but also an efficient, safe, comfortable, economic, low-carbon, and environment-protecting transportation sys-tem. The SSR concept is not just an idea now; it has developed into a technological revolution to promote social progress and human development.

Especially with the improvement of living standards, people no lon-ger care only about whether they can reach their destination quickly; the travel process has become an important part as well. The SSR system will not simply optimize people's travel experience but also completely inno-vate it. Therefore, the construction goal of the SSR system is not only to surpass the aircraft in speed but also to create an all-round new travel experience, surpass the aircraft in comfort and convenience, and bring people a more comfortable and pleasant travel experience.

Bibliography

A. S. Abdelrahman, J. Sayeed, and M. Z. Youssef. Hyperloop transportation system: Analysis, design, control, and implementation. *IEEE Transactions on Industrial Electronics* 2017; 65(9): 7427–7436.

A. C. Adoko and Li Wu. Estimation of convergence of a high-speed railway tunnel in weak rocks using an adaptive neuro-fuzzy inference system (ANFIS) approach. *Journal of Rock Mechanics and Geotechnical Engineering* 2012; 4(1):11–18.

J. Braun, J. Sousa, and C. Pekardan. Aerodynamic design and analysis of the Hyperloop. *AIAA Journal* 2017; 55(7):1–8.

J. Chai, Y. Zhou, X. Zhou, *et al.* Analysis on shock effect of China's high-speed railway on aviation transport. *Transportation Research Part A: Policy and Practice* February 2018; 108:35–44.

Z. Dajin. *Research on SS-HTS Maglev Train System with High Speed Annular Pipeline*. PhD dissertation, Southwest Jiaotong University, 2017.

S. Ya. Davydov, N. P. Kosyrev, N. G. Valiev, *et al.* Theoretical studies of the unloading of containers in the pneumatic transport systems of today and tomorrow. *Refractories and Industrial Ceramics* July 2013; 54(3):178–187.

B. Eom, C. Lee, S. Kim, *et al.* Investigation on prototype superconducting linear synchronous motor (LSM) for 600-km/h wheel-type railway. *Physics Procedia* 2015; 65:261–264.

S. Fabozzi, E. Bilotta, M. Picozzi, *et al.* Feasibility study of a loss-driven earthquake early warning and rapid response systems for tunnels of the Italian high-speed railway network. *Soil Dynamics and Earthquake Engineering* September 2018; 112:232–242.

Y. Gao, H. Huang, C. L. Ho, *et al.* High speed railway track dynamic behavior near critical speed. *Soil Dynamics and Earthquake Engineering* October 2017; 101:285–294.

M. Gao, Y. Wang, Y. Wang, *et al.* Experimental investigation of non-linear multistable electromagnetic-induction energy harvesting mechanism by magnetic levitation oscillation. *Applied Energy* June 2018; 220:856–875.

K. V. Goeverden, D. Milakis, M. Janic, *et al.* Analysis and modelling of performances of the HL (Hyperloop) transport system. *European Transport Research Review* 2018; 10(2):1–17.

W. Haiyang. *Aerodynamic Characteristics and Energy Consumption Analysis of Evacuated Tube Transport System at High Speed.* Hunan: Hunan University, 2018.

J. Hitt. Ready for takeoff. *Smithsonian* 2016; 47(2):88–96.

A. Hyde-Wright, B. Graham, and K. Nordback. Counting bicyclists with pneumatic tube counters on shared roadways. *Institute of Transportation Engineers Journal* March 2014; 84(2):32–37.

R. Janzen. TransPod ultra-high-speed tube transportation: Dynamics of trains and infrastructure. *Procedia Engineering* 2017; 199:8–17.

G. Jiangmin. Second kill maglev vehicle — SSR. *China Economic Report* 2016; 18(6):114–116.

T. Kashiwagi, M. Kubota, E. Suzuki, *et al.* Levitation characteristics of the superconducting mixed-μ system. *Physica C: Superconductivity* October 2003; 392(3):654–658.

J. S. Lee and J. H. Kim. Approximate optimization of high-speed train nose shape for reducing micropressure wave. *Structural and Multidisciplinary Optimization* 2008; 35(1):79–87.

W. Lei. *Research on Bearing Pipeline of Vacuum Pipeline Maglev Train.* Chengdu: Southwest Jiaotong University, 2011.

W. Y. Liu, J. G. Han, and X. N. Lu. A high speed railway control system based on the fuzzy control method. *Expert Systems with Applications* 1 November 2013; 40(15):6115–6124.

F. Liu, S. Yao, J. Zhang, *et al.* Effect of increased linings on micro-pressure waves in a high-speed railway tunnel. *Tunnelling and Underground Space Technology* 2016; 52:62–70.

F. Liu, S. Yao, J. Zhang, *et al.* Field measurements of aerodynamic pressures in high-speed railway tunnels. *Tunnelling and Underground Space Technology* 2018; 72:97–106.

G. R. Mullins and D. E. Bruns. Air bubbles and hemolysis of blood samples during transport by pneumatic tube systems. *Clinica Chimica Acta* October 2017; 473:9–13.

D. Oster, M. Kumada, and Y. P. Zhang. Evacuated tube transport technologies (ET3)[tm]: A maximum value global transportation network for passengers and cargo. *Journal of Modern Transportation* 2011; 19(3):42–50.

Z. Peihao. Can the "SSR" start smoothly? People's Weekly 2016; 8(1):46–47.

H. Qizhou, L. Xianghong, and Q. Siyuan. *Brief History of High Speed Rail.* Chengdu: Southwest Jiaotong University Press, 2018.

S. L. Quellec, M. Paris, C. Nougier, *et al.* Pre-analytical effects of pneumatic tube system transport on routine haematology and coagulation tests, global coagulation assays and platelet function assays. *Thrombosis Research* May 2017; 153:7–13.

Md. A. Sayeed and M. A. Shahin. Three-dimensional numerical modelling of ballasted railway track foundations for high-speed trains with special reference to critical speed. *Transportation Geotechnics* March 2016; 6:55–65.

S. Thalén, I. Forsling, J. Eintrei, *et al.* Pneumatic tube transport affects platelet function measured by multiplate electrode aggregometry. *Thrombosis Research* July 2013; 132(1):77–80.

C. Thompson. The next pipe dream. *Smithsonian* 2015; 46(4):17–18, 20, 23.

S. Türkay and A. S. Leblebici. Effect of multi-objective control on ride quality in high speed railway train. *IFAC-PapersOnLine* 2016; 49(3):273–278.

Y. Wang, W. Ren, J. He, *et al.* Analysis of aerodynamic loading properties on hood of high-speed railway tunnel. *Perspectives in Science* March 2016; 7:323–328.

C. Xi. SSR leads the new speed in the future. *Modern Industrial Economy and Informatization* 2013; 17(13):80–81.

L. Xia. Favorite of future traffic stars. *Knowledge is Power* 2016; 23(4):30–35.

C. Xuyong. *Simulation Analysis of Aerodynamic Problems of Vacuum Pipeline Maglev Train.* Chengdu: Southwest Jiaotong University, 2010.

T. Youfu. Analysis on the development trend and key problems of SSR. *Railway Construction Technology* 2019; 59(4):1–4.

Y. P. Zhang, D. Oster, M. KtLmada, *et al.* Key vacuum technologies to be solved in evacuated tube transportation. *Journal of Modern Transportation* 2011; 19(2):110–113.

Y. Zhao, H. He, and P. Li. Key techniques for the construction of high-speed railway large-section loess tunnels. *Engineering* April 2018; 4(2):254–259.

H. Zhu and J. Yang. Modeling and optimization for pneumatically pitch-interconnected suspensions of a train. *Journal of Sound and Vibration* October 2018; 432:290–309.

Index

www.ingramcontent.com/pod-product-compliance
Lightning Source LLC
Chambersburg PA
CBHW050640190326

41458CB00008B/2348